D1015895

THE LAST HARVEST

The Genetic Gamble That Threatens to Destroy American Agriculture

PAUL RAEBURN

Simon & Schuster

New York London Toronto Sydney Tokyo Singapore

SIMON & SCHUSTER
Rockefeller Center
1230 Avenue of the Americas
New York, NY 10020

Designed by Liney Li
Manufactured in the United States of America

10 9 8 7 6 5 4 3 2 1

Library of Congress Cataloging-in-Publication Data
Raeburn, Paul.
The last harvest: the genetic gamble that threatens to destroy American agriculture / Paul Raeburn.
p. cm.
Includes bibliographical references (p.) and index.
1. Plant breeding—United States. 2. Crops—United States—Germplasm resources. 3. Agricultural innovations—United States. 4. Agriculture—United States. I. Title.
SB123.25.U6R34 1995
333.95'3—dc20 95-7369
ISBN 0-684-80365-8

TO MY PARENTS

What would the world be, once bereft
Of Wet and Wildness? Let them be left,
O let them be left, wildness and wet;
Long live the weeds and the wilderness yet.

—*Gerard Manley Hopkins*

Contents

Introduction

Betting the Farm

In June 1988, I was assigned by the national desk of the Associated Press in New York to cover a conference at the headquarters of the federal government's Agricultural Research Service in Beltsville, Maryland. The conference was entitled "Biotic Diversity and Germplasm Preservation: Global Imperatives." With little time to prepare, I took the shuttle to Washington, walked into the auditorium where the conference was being held, and took a seat.

When I arrived, a man I later learned was Garrison Wilkes, a botanist at the University of Massachusetts in Boston, was speaking in apocalyptic terms about "genetic erosion" and "seed morgues." Some kind of crisis was brewing, and Wilkes was worried about it. But I had trouble discovering what that crisis was. Wilkes was not talking about endangered species, which I would have understood. He was talking about corn and wheat.

As he spoke, heads nodded in agreement in the audience—what little audience there was. The Agricultural Research Service auditorium was less than half full. How serious could the crisis be if it could not draw a full house?

With a deadline to meet, I pulled Wilkes aside at the first opportunity and asked for help. He told me about something called southern corn leaf blight. In 1970, it had wiped out 15 percent of the nation's corn crop at a cost of $1 billion. The losses could easily have been even higher, Wilkes said. This time, we were lucky.

The problem, Wilkes explained, was that the same thing could happen again. And the next time it might be worse.

Each of the millions of cornstalks that springs from the rich soil of the U.S. corn belt is nearly an identical twin of every other one. You do not have to be an expert to see it. The corn plants are of identical height and shape. The ears are indistinguishable. Each has the same genes for leaf size and stiffness of stem.

By focusing on a handful of the very best varieties, grain-belt farmers continue to squeeze ever better harvests from their fields. But there is a downside to this concentration on the same few corn varieties. Never before have so many millions of acres around the world been covered by plants that are nearly identical.

American farmers, part of one of the most productive agricultural enterprises in history, are waging a genetic gamble. They are "betting the farm" in a way they never intended. Many are not even aware of the risks they face.

It was that unwitting gamble that concerned Wilkes. We, as consumers, are partly responsible. We want the largest, sweetest ears of corn, the tastiest tomatoes, and the juiciest oranges. Supermarkets display mountains of uniformly attractive fruits and vegetables. We expect nothing less.

One reason why American farmers have been so successful is that American crops are highly resistant to predators. The crops have been bred that way. But resistance is a fleeting thing. It is a game of leapfrog between breeders and predators. Breeders produce a new crop variety that resists common pests and diseases. Within a few years, an insect or microbe mutates. It acquires the ability to overcome the resistance. It develops a taste for a bitter leaf. It learns to burrow through a toughened stem. It finds a new route of infection. Breeders respond with a variety that is resistant to the new predator. Farmers gain a few more years. Then the cycle repeats. Each variety has an average lifetime of about seven years.

Genetic uniformity raises the stakes in that game of leapfrog. If a pest finds one cornstalk tasty, it will find them all tasty, all across the Midwest.

Admittedly, the possibility of a catastrophic collapse of the grain belt is hard to accept. American farmers easily feed 250 million Americans and much of the rest of the world as well. The United States is the world's leading producer of crops for export. In 1991, American farmers grew 66 percent of the world's export corn, 32 percent of its export wheat, and 66 percent of its export soybeans.[1] That doesn't sound like an enterprise in danger.

By betting that no pests or disease will come along to take advantage of the uniformity in their crops, farmers are playing long odds. Nature is predictable. Pests and diseases will leapfrog ahead of breeders. They will exact their biological revenge. Seed companies and farmers ought to take a lesson from Las Vegas. You can keep a winning streak going for a long time. But in the end, the house always wins.

A few farsighted biologists anticipated this threat at the turn of the century. They began sending collecting expeditions around the world in search of seeds of potentially valuable wild and traditional crop varieties. The largest collection of these seeds is run by the U.S. Department of Agriculture and housed in dozens of cavernous refrigerated vaults across the country. The flagship of the U.S. "seed banks," as they are called, is the National Seed Storage Laboratory (NSSL) in Fort Collins, Colorado. More than a quarter of a million packets of seeds are stored there. In 1991, a panel assembled by the National Academy of Sciences called the U.S. seed collections "a strategic resource essential to national and global agricultural security."[2]

The seed banks hold the raw materials for broadening the genetic diversity in American crops. Refrigerated vaults hold hundreds of thousands of seeds as different from one another as the multicolored kernels on an ear of Indian corn. As Wilkes explained to me, however, the seed banks are not providing the security they are supposed to. They, too, are in trouble.

The stakes are growing higher as farmers now contend not only with the threat of disease but with changes in the environment around them. Global warming, the depletion of the ozone layer, and polluted air are new threats to farmers' crops.

"To keep every cog and wheel is the first precaution of intelligent tinkering," wrote the naturalist Aldo Leopold.[3] If the genetic cogs and wheels of agriculture are lost, American crops will become even more vulnerable to pests, diseases, and global change. There are many reasons why we might want to preserve what remains of the world's wilderness. This book adds another one: Without the biological resources to keep improving our crops, we could one day go hungry.

The Last Harvest

Chapter 1

Billion-Dollar Corn

Hugh Iltis, a University of Wisconsin botanist, was in a jeep on a dusty road in southeastern Guatemala when he saw a disturbance ahead. Several armed militiamen had established a roadblock. Iltis, who was on a botanical collecting expedition, approached cautiously. The militiamen grabbed their weapons. One of them stopped Iltis's jeep. "He told me to get out and started yelling in Spanish," Iltis recalled. "And he was especially yelling at my guide." The guide tried to talk to them. He was careful not to make any quick movements or to threaten them. "They were scary, because it was miles from noplace," Iltis said. He had been prospecting in the countryside, and everything he had collected was plainly visible on the backseat inside the jeep. The cargo was irreplaceable. Iltis was determined to protect it.

Summoning experience gained from a lifetime of traveling to remote corners of Latin America, he seized the initiative. Shouting in what he calls Spanglish, he upbraided the men for getting in his way. "I finally pulled out my documents, with a gold seal, and started yelling at them that this was a *misión oficial.*" As he customarily does before leaving for Mexico, Iltis had asked university administrators to give him a formal letter of introduction, complete with a university seal. "I never cracked a smile. You never want to crack a smile with these people. I said they better watch out. They told my guide he better get lost. I said, 'He is staying with me. I'm going to call the governor. You are going to end up in real trouble.' They looked at the letter, which looked very official."

For agonizing minutes, the leader of the group considered Iltis's letter. It was in English, which he could not read. Iltis pointed to the seal. "Finally, he backed off, grumbled something, and told the guide to get back in the car. And off we drove." Iltis's precious cargo was secure.

A broad-shouldered sixty-nine-year-old with thick white hair and a gray brush mustache, Iltis is one of a small band of botanist-explorers who devote their lives to scouring the world for rare botanical treasures. The aim of these explorers, whose work goes largely unnoticed, is to find genetic breeding stock that can be used to improve farmers' crops. Their contributions to agriculture have been immense, helping to feed billions who would otherwise have starved to death. High-yielding crop varieties—developed partly with the help of plants like those discovered by Iltis and his colleagues—made a major contribution to the explosive growth in food production during the twentieth century. Iltis himself was responsible for a discovery that could revolutionize the cultivation of corn in the U.S. grain belt and around the world. The discovery has been called the botanical find of the century.

Iltis has the energy, enthusiasm, and booming voice of a much younger man. He freely offers tough, blustery denunciations of those with whom he disagrees—which includes almost everyone—on some point or other. A favorite Iltis target is the pope, for his opposition to birth control. Iltis's criticism of the pope wins him few sympathizers, but he defiantly persists in raising the subject at every opportunity. In front of a Catholic audience, he is certain to do it.

Iltis softens, however, when he talks of the history and evolution of a rare flower. The fascination with nature that led him to a career in science suddenly becomes evident. With his weathered looks, salty language, and brusque manner, he lacks the demeanor of an academic. At the University of Wisconsin, where he has spent most of his career, Iltis is an emeritus professor of botany and the director of the university's herbarium. From his base in Madison, Iltis has made numerous trips to Latin America, most of them to Central America and Mexico, in search of rare plants.

The cargo he had risked his life to protect in Guatemala was a stack of plant presses stuffed with a tough, stringy plant called teosinte, a wild relative of corn. The militiamen had no interest in the teosinte. They were after Iltis's jeep. If they had noticed the teosinte, they probably would not have recognized its value. Iltis was afraid they might discard it out of suspicion or spite. Almost certainly they would have dismissed Iltis as some sort of misguided eccentric.

The militiamen were not alone in their failure to recognize the value of Iltis's cargo. Few scientists would have appreciated what Iltis was doing. To anyone but Iltis or a few dozen other botanists, the plants looked like worthless kindling. The scrubby teosinte vaguely resembles a corn plant. But instead of bright buttery yellow ears of corn, it bears measly, finger-sized ears, each with only a few tough, misshapen kernels. Ears of teosinte would never be mistaken for cultivated corn.

Iltis was collecting the teosinte for two reasons. One was to help fill in the blanks in corn's complicated and poorly understood family tree. Another was to search the teosinte for genes carrying pest and disease resistance that might be crossbred into cultivated corn, to help it withstand nature's attacks in the field. The teosinte he had gathered might be useless itself as a commercial crop; but its seeds, with their tangle of genetic material—what botanists refer to as *germplasm*—almost certainly would carry some interesting disease-resistance genes.

Corn is the largest and most important crop grown in the United States. As Noel Vietmeyer at the National Academy of Sciences has pointed out, corn does much more than feed America. "You may rely on corn products each time you read a magazine, walk across a carpet, mail a letter [cornstarch is used as sizing on textiles and paper and as glue on stamps], dine on a steak, take a beer, a soft drink or a whiskey, eat bread or candy, chew gum, light a match, or take an aspirin or penicillin."[1]

Along with its strengths, however, corn has a fatal weakness. It cannot survive without a farmer's steady hand. Indeed, corn cannot reproduce without human help. The tightly wrapped husk that makes corn easy to harvest prevents the corn kernels—the seeds—from escaping and reaching the soil. Without human hands to remove the husk, no corn plant can give rise to offspring. For centuries corn has been bred for yield and quality, not resilience. It is vulnerable to pests and diseases, and it requires a hefty dose of nutrients. Teosinte, on the other hand, is wild. It flourishes in the absence of pesticides, herbicides, or human assistance.

In order to have survived in Guatemala's simmering tropical cauldron of bugs and microbes, the plants Iltis had collected needed their own built-in disease and pest resistance. Such resistance is precisely what breeders must have if they are to win the constant battle between pests and cultivated crops. If Iltis's teosinte specimens contained even a few new disease-resistance genes, they would be enormously valuable to American farmers—far more so than anything the Guatemalan militiamen might have been seeking.[2]

Iltis had an auspicious beginning for a botanist. He was born in 1925, in Brno, Czechoslovakia. The city, then part of Austria-Hungary, was known in German as Brünn. It was there, nearly a century earlier, that the great botanist Gregor Mendel discovered the laws of modern genetics by crossbreeding pea plants.

Iltis's father, Hugo, was eighteen in 1900 when Mendel's research was rediscovered after more than fifty years of neglect. The elder Iltis met many of the famous people who came to Brünn to see the monastery where Mendel had worked. Iltis's father became a botanist himself—a professor of plant morphology, the study of plant structure. In 1911 he wrote a paper on the evolution of corn, a subject that would occupy his son fifty years later. He also became Mendel's biographer. Near the end of World War II, Iltis's mother discovered—in the course of proving to German authorities that she was Aryan—that Mendel was her great-great-great-great-uncle.

"I grew up in a family with a lot of natural history—white mice and lizards in cages, and gardening, and especially field trips, hikes," Iltis recalled. "We were schooled in natural history quite determinedly. If I didn't know the name of a plant and my father got impatient, I got slapped." By the time Iltis came to the United States, when he was thirteen, he had already been collecting plants for five or six years.

Iltis's father was Jewish. He attracted the attention of the Nazis for his writings against racism, including a book called *The Myth of Blood and Race.* (After the war, Iltis, then an American intelligence officer, entered a Nazi party headquarters and found stacks of his father's books.) Iltis's father became despondent over the course of the war and attempted suicide in 1938. A short time later, after weighing offers of professorships in Moscow and the United States, the senior Iltis left for a teaching job in Fredericksburg, Virginia. Hugh and his mother followed in 1939.

In the United States, Iltis resumed his fieldwork, collecting plants, frogs' eggs, and butterflies and establishing a herbarium. He became friendly with curators at the Smithsonian Institution in Washington. In January 1944, at the age of eighteen, he found himself on a hundred-ton fishing trawler off Cape Hatteras collecting fish for the Smithsonian. He wrote and published his first scientific paper later that year, before going into the army. He arrived in Europe the week the war ended. Because he spoke German fluently, he was put to work interrogating captured German officers and examining documents stored by the Nazis in Heidelberg. The statements he obtained were used to prosecute German war criminals.

After his military service, Iltis took a job at the University of Tennessee. He chose it for its proximity to the Smoky Mountains. "I was in love with the Smoky Mountains," he said. After graduation, he moved to the Missouri Botanical Garden, where he earned a Ph.D. in botany. While he was there, he made a trip to Costa Rica, his first visit to Latin America. On that trip he discovered a new genus of daisy, now known as *Iltisia.*

He also saw, for the first time, the widespread destruction of tropical forests. He immediately wondered what he might do to stop it. "I was very much interested in preservation," he said. He spent a few years at the University of Arkansas, moved to the University of Wisconsin in 1955, and returned to Mexico in 1960. For the first time, he said, he "really saw the damage." He visited an area where earlier visitors had reported seeing a population of *Zea perennis,* a relative of corn. Botanists feared it had become extinct in the wild. Although it had been reported in the area Iltis visited, he could find no trace of it.

In 1962, Iltis took another trip, this time to Peru. The outcome was notably more successful. Iltis stumbled upon a botanists' El Dorado, bringing back a handful of seeds that have transformed the California tomato industry.

Iltis made the trip with Donald Ugent, now a professor at Southern Illinois University at Carbondale. They went in search of wild potatoes. For a month they collected wild potatoes in the hills above Lima, trying to determine how the wild varieties might have evolved into something resembling modern, cultivated potatoes. Then they followed a rough gravel road along the front range of the Andes up to a vast, arid, tundralike grass plain called the *puna.* From there the road coiled into a series of steep switchbacks as it descended six thousand feet to the golden city of Cuzco, once the seat of the Inca kingdom.

At the end of the month, Iltis organized and packed fifteen hundred specimens of three hundred different species of plants he and Ugent had collected. Meanwhile, Ugent went to a nearby valley. There he discovered a population of yet another rare wild potato. "That was enough to make us stay one more day, to revisit that valley together, study its flora, and scour its slopes for potatoes," Iltis said. On their way, the researchers were stopped by a landslide. They left their jeep, hiked along the road, and came upon a valley Iltis described as "a floristic wonderland full of rare and beautiful plants." Spanish moss hung from cliffs and tree branches. The ground was carpeted in wildflowers and giant cacti, and iridescent blue-and-green hummingbirds darted back and forth. They stopped for lunch.

At their feet, Iltis and Ugent spotted "a tangled, yellow-flowered, sticky-leaved, rather ratty-looking wild tomato." It was not much different from other tomatoes they had seen. (Tomatoes and potatoes are members of the same plant family, and wild varieties of both are found in Peru.) The plant bore green-and-white-striped "tomatoes" smaller than cherries. Iltis and Ugent sent some of the plant's seeds to Charles Rick, a geneticist at the University of California, Davis, probably the most famous tomato geneticist in the world. Rick wrote back thanking them, saying that they had apparently found a new species.

"Our story could have ended here, of course, and still be a good one, what with us showing off the type of specimens housed in the University of Wisconsin Herbarium to interested students and telling tall expedition tales of haciendas and vicuñas, potatoes and tomatoes," said Iltis. "But there was more to come."

Rick began serious work on the tomatoes, planting the seeds, studying the seedlings, and crossbreeding them with cultivated tomatoes. It was not until 1980, more than seventeen years later, that Rick told Iltis what had become of his weedy tomato. When Rick crossed the new species with commercial tomato varieties, he found that the progeny had a higher percentage of soluble solids—mostly sugars. That is precisely the characteristic tomato canners and processors are looking for. The higher the percentage of soluble solids, the more valuable the tomatoes.

Commercial tomatoes were about 4.5–6.2 percent soluble solids. Using Iltis's tomato, Rick produced tomatoes with as much as 8.6 percent soluble solids—an increase of roughly 40 percent. The tomato discovered by Iltis and Ugent was truly a botanical gold mine, and it was quickly incorporated into commercial tomatoes. It has increased the value of the California tomato crop by $21 million a year.

That is not a bad return on the money the National Science Foundation invested in the expedition that sent Iltis and Ugent to Peru. The yearlong trip and three years of follow-up research cost the federal government $21,000. It yielded a thousand herbarium specimens, at a cost of $21 per specimen. That $21 million a year in profits for the tomato industry is the return on a $21 wild tomato.

Even that may not be the last word on the Iltis-Ugent expedition. "Perhaps the most significant values stemming from our expedition are yet to come, possibly from the high-protein potatoes we collected or from the hundreds of bits and pieces of botanical information we passed along to colleagues, graduate students and others," Iltis wrote later. "But as in the case of our tomato, collected in 1962, commercially utilized a decade later, and not described as a new species until 1976, the practical value of an organism can often not be recognized except after years of work."[3]

At about the same time he and Ugent found the tomato, Iltis began studying grasses and corn. (Corn, wheat, rice, and other grains are members of the grass family.) By the 1970s, he had become part of what are sometimes called "the corn wars," a continuing academic feud over one of the great puzzles in botany: the origin of corn. Despite its ubiquity, no one knows where cultivated corn came from. There is no clear ancestral path that shows what its wild progenitors were or how they gave rise to it.

Iltis was one of an all-star team of fifteen American botanists who went to Mexico in 1971 to look for specimens of teosinte that might shed light on the question. The expedition failed to end the corn wars. But it had a great effect on Iltis's career. "We were in a group collecting teosinte seeds, looking for mutations. We didn't find any, but that was a turning point for me, because I'd never seen teosinte before in the wild."

Iltis was fascinated by teosinte. That fascination ultimately led him on a treasure hunt that ended with the discovery often referred to as the botanical find of the century. Iltis's teosinte discovery as well as his weedy tomato provide dramatic illustrations of the potential value of the germplasm in wild plants.

The teosinte discovery might never have been made if not for Iltis's unusual habit of sending hand-drawn botanical New Year's cards to friends and colleagues. In December 1975, Iltis's card read: "Peace to all Mankind, Goodwill to the Earth, To all its flowers, birds and children, both young and old, and to you—a Happy New Year, 1976." More important than the greeting was the illustration on the card. Iltis had decorated it with his drawing of the plant he had been unable to find in Mexico in 1960: *Zea perennis.* The plant hadn't disappeared; seeds were kept by researchers around the world. But as far as anyone knew, all natural populations were gone. Underneath the drawing on the card, Iltis wrote, "Extinct in the Wild."

Iltis sent the card to associates all over the world. He had little reason to think that anyone would take up the implicit challenge to prove him wrong by searching for *Zea perennis* in the wild. As it turned out, however, one of the hundreds of cards that Iltis mailed landed on fertile ground, like a lone seed that chances to find a pocket of soil on a rocky outcrop.

One of the recipients of Iltis's card was Luz María Puga, a professor of botany at the University of Guadalajara. She was skeptical of Iltis's assertion that *Zea perennis* was extinct in the wild. There was always the possibility that a few plants had escaped detection and were thriving, unrecognized, along a roadside somewhere or in a neglected corner of a farmer's field.

She said nothing to Iltis, but she passed the card along to one of her students, a young man named Rafael Guzmán. Guzmán was a capable student who had made a virtue of hard work and persistence. Iltis thinks this species is extinct, she told Guzmán. Do you think you can prove him wrong?

Iltis had talked to Puga four years earlier about searching for wild populations of this particular species of teosinte. Its scientific name reflected its unique characteristic: Unlike corn, *Zea perennis* survived the Mexican winter. It was a perennial. No other perennial form of teosinte was known. Iltis was intrigued by it. The interest was primarily scientific. *Zea perennis* had twice as many chromosomes as cultivated corn and thus could not easily be bred with corn. It appeared to have little practical value. Nevertheless, Iltis is not the type to let an unresolved scientific question remain unresolved. The greeting card was his attempt to fire up some enthusiasm for a better search.

Guzmán had been at the University of Guadalajara since 1971, studying the classification of grasses. When Puga gave him Iltis's card, Guzmán pulled out a topographical map of Mexico and looked for locations where the conditions were right for *Zea perennis* to thrive. Unfortunately, there were many. He marked five hundred of them—far too many to explore. A less determined researcher might have abandoned the project right there. But not Guzmán. Of the five hundred, he arbitrarily selected five.

"I went to the first one and didn't find anything," Guzmán said. The second location turned up nothing, as did the third and fourth. Guzmán carried Iltis's greeting card with him to each spot. As he traveled to and from the locations he had selected, he would stop the local farmers, or *campesinos,* show them Iltis's card, and ask them whether they had seen anything like it.

Guzmán was searching in cornfields near Colima, south of Guadalajara, when a passing campesino asked Guzmán what he was doing. Guzmán showed him the greeting card. "He said he knew this plant," Guzmán said. The campesino said he had seen it growing in a place he knew not far away. Come back in three hours and I will take you there, the campesino told Guzmán.

"Later the same day, he showed it to me," Guzmán said. That was in mid-August of 1977. The place to which the campesino took Guzmán was, coincidentally, the fifth and last location on Guzmán's list.

Guzmán was not sure it was the plant he was looking for. The campesino insisted that it was. Guzmán dug it up, took it back to the greenhouse at the University of Guadalajara, and planted it. He waited and watched, finally concluding that the campesino was right. It was indeed *Zea perennis.* "If he hadn't taken me, I wouldn't have recognized it," Guzmán said. But that was not yet the end of the story.

When Guzmán's plant flowered in the University of Guadalajara greenhouse, another student told Guzmán that he had seen it before—in his grandfather's *huerto,* or orchard. The student gave Guzmán directions, and Guzmán set out to discover whether he could find another population of the plant that was supposed to have been extinct in the wild. He hiked for ten hours over steep ridges in the Sierra de Manantlán to reach the *huerto,* only to discover that the student had been mistaken. All Guzmán found on the farm was a form of corn, the *parviglumis* subspecies of *Zea maize.* It was an annual.

Again, however, Guzmán was lucky. A farmer passing on a mule asked Guzmán what he was looking for. When Guzmán told him, the farmer said he knew of a perennial form of that plant in a place twelve kilometers farther into the mountains. Pointing to a small path, the farmer told Guzmán to take it to the village of El Durazno, at the base of a mountain. Sleep there, the farmer said, and the next day, go up the mountain to La Ventana. Then go on to San Miguel and you will find lots of this.

The indefatigable Guzmán continued his trek. This time he was rewarded. He found hundreds of perennial teosinte plants. He could tell they were perennials by the presence of root structures called rhizomes, which allow the plants to survive the winter. In 1979, he recounted his discoveries in two scientific papers. One reported the discovery of *Zea perennis.* Another reported the discovery he had made along the way of a population of a related species, *Zea mexicana.*

Iltis, in his herbarium in Madison, read the papers with great excitement. As it happened, they appeared just after Iltis had received a curious letter from a New York City man named Tony Pizzatti. The letter said, simply, "Friends of mine found the ancestor of corn." Pizzatti had taken his letter to the New York Botanical Garden, which had referred him to Iltis. Not knowing what to make of the letter, Iltis wrote back, asking for more information. "The next letter came, and he said they found something called *perennis,* or something like that," Iltis said. "Oh, boy, I still remember. I said, 'It can't be.' This guy tells me they found *Zea perennis,* lost since 1921."

Iltis and one of his students, John Doebley, now at the University of Minnesota, immediately wrote to Guzmán. "We said, 'Please send us specimens, seeds, anything.' He said, 'I have no specimens; all I have is seeds.' " Guzmán sent Iltis some of the seeds. Doebley stuck them in a pot and put it on his windowsill.

At the end of June, Doebley walked into Iltis's office with the pot. "He said, 'Here it is, it's blooming.' A little scrawny plant. It was supposed to be *perennis.* I thought, Geez, I don't know. That doesn't look like anything that I've ever seen. The leaves were too narrow, for one. It was peculiar, but partly because it was grown in a pot. It could have been *perennis,* but I just doubted it."

At the time, a woman from Hebrew University named Batia Pazy was working with Iltis in Madison, analyzing chromosomes in plants in the caper family. When Doebley came to Iltis with the flowerpot from his windowsill, Pazy was finishing the work and preparing to return to Israel. "I said, 'Batia, for God's sake, help

John make some squashes,' as they're called—squash the pollen grains and figure out what the chromosome number is." Pazy agreed, and she went to work, while Iltis left for northern Wisconsin on a collecting expedition with a group of visiting Russian scientists.

Zea perennis has forty chromosomes. Cultivated corn has twenty. Iltis wanted to confirm that the plant in Doebley's flowerpot had forty chromosomes. When Iltis returned, Pazy and Doebley were standing on a balcony overlooking the lobby outside his office. "I can still see them saying, 'Well, we've got news for you. We don't know what the chromosome number is, but it seems closer to twenty than forty.' I looked at them, and my mouth dropped open, and the wheels were going, and I said, 'Really? Are you sure?' And they said, 'Oh, yeah, no question about it.' I said, 'Do you know what that means? We've got a new species.' "

Pazy soon confirmed that the plant that had bloomed on Doebley's windowsill had exactly twenty chromosomes. That was critically important. Corn has twenty chromosomes. Whatever this thing was, it could be crossbred with corn. Pazy, Doebley, and Iltis knew immediately what that could mean: It was now possible for breeders to create perennial corn.

Such an achievement could completely transform the American grain belt. American farmers grow corn on 60 million acres. Each year, farmers move row by row across those 60 million acres of topsoil to plant corn. The corn is watered and tended over that vast acreage. Pesticides and herbicides are applied where they are needed. Then, at harvesttime, farmers move across the entire 60 million acres again. The corn is carried away, and the fields are prepared for the next planting.

Iltis and Guzmán's discovery could dramatically revise that picture. The grain belt could theoretically be transformed into a 60-million-acre corn "orchard." Farmers would plant corn "trees" once. The perennial corn would yield a harvest year after year, without annual planting, without extensive preparation—and consequent erosion—of the soil.

Geneticists who had done evolutionary studies of teosinte had predicted that there must be a perennial relative of corn with twenty chromosomes. Yet no one had ever found it. "Well, there it was," said Iltis. He and Doebley named the new species *Zea diploperennis.* And they immediately realized that they had a problem.

It was July, too early to obtain additional live, blooming specimens in Mexico. "We had two choices," Iltis said. "We either could describe it right away or go and get good material, because that material wasn't any good in the pot." If they waited, they could prepare a much more comprehensive report, using specimens that had developed under natural conditions. But that would also mean taking a chance that their discovery would somehow leak out, that it would be whisked away from them. When it comes to scientific discovery, there is no second place. Either they would be remembered for finding it, or someone else would. They decided to take the risk and go to Mexico.

"We realized almost instantly that an expedition had to be launched to collect living and preserved material in the field, for we knew next to nothing about the taxonomy, ecology and geography of this new species," Iltis said. "And any relative of the world's third most important food plant [corn] is *ipso facto* of supreme interest to biology and agriculture."[4]

Iltis and Doebley drove from Wisconsin to Mexico, arriving in Guadalajara on about September 15. The next day, they went looking for Guzmán, who, they mistakenly assumed, was a professor at the University of Guadalajara. They could not find him; there was no record of a Dr. Guzmán. But they persisted. "Finally, this young kid, twenty-two years old, said, grinning, 'I'm Rafael Guzmán.' The director of the botanical institute was very unhappy, because he thought we wanted to talk to him," Iltis said. "All we wanted to do was to talk to Guzmán."

Iltis and Doebley asked Luz María Puga to join them. Iltis told Guzmán that what he found was not *Zea perennis.* "His face just fell a mile," Iltis said. Then Iltis shared the secret. "Within the first hour in the Instituto de Botanico at the University of Guadalajara, we informed both Guzmán and Puga of the fact only known to us: namely, that the Sierra de Manantlán population was an undescribed new species of great biological and practical interest, an announcement which surprised and pleased them both immensely."

It took Iltis and Doebley six days to get near the place where Guzmán had found the seeds that Doebley had planted in the flowerpot back in Wisconsin. "The next day, we got up very early. We headed out at six in the morning, through magnolia forests and cloud forests of oak. The view was just magnificent," Iltis said.

At one point, there was a brief scare. "I remember riding ahead and looking up, and there were three guys on horses with big white sombreros and guns across the saddle, standing there like statues, not moving a muscle." They watched Iltis for a while, came toward him on the trail, passed him—and disappeared.

By the end of the day, Iltis, Doebley, and Guzmán had reached the spot where Guzmán had found the seeds. "There was lots of it," Iltis said. "It was clearly perennial teosinte. It looked like it. But then I dug it out, and it was very different. It had very short, fat rhizomes, about the size of radishes, small radishes. Very different. *Perennis* has long rhizomes, a foot to two feet long, about the size of a pencil or a thick ballpoint pen, with knots in it."

Guzmán washed off the roots, and Iltis confirmed that they were looking at a new species. "I held it up and I said, 'Oh, boy, man alive, there's no question.' We stayed there about three hours, collecting everything we could, including twenty-five specimens of that grass for duplicates."

They returned to Guadalajara, where they had a victory dinner a few days later. While they were eating, the phone rang. It was a botanist from Mexico who had heard about the discovery of *Zea perennis.* He said he had a hunch about what it might be and that he would arrive in Guadalajara in three weeks. "The race was on," Iltis said.

In November 1978, Iltis, Doebley, and Guzmán prepared a scientific report for *Science,* one of the most prestigious American scientific journals. The paper was published the following January, certifying their discovery of the long-sought *Zea diploperennis.* (The name reflects the fact that it has half as many chromosomes as *Zea perennis.*)

The discovery was widely reported in newspapers and on television. "It changed my life," Iltis said. "All hell broke loose." He was swamped with requests for more information and demands for seeds. "Telegrams came, [saying,] 'We want to order four hundred pounds of seeds of this thing.' You know how much we had? About two ounces."

The excitement was fueled by a powerful image: of perennial corn. Iltis encouraged the media interest by slyly concluding in the *Science* paper that "this new species . . . may lead to the development of perennial corn."

It was a bit of a false hope, and Iltis knew it. It is unlikely that a perennial hybrid of *Zea diploperennis* and commercial corn will ever completely take over the U.S. grain belt. One reason is that *Zea diploperennis* will not survive northern winters. So corn bred from it might not fare well in the northern United States, either. But the climate in the Sierra de Manantlán mountains, where *Zea diploperennis* was found, is similar to that of the southern United States. Perennial corn might survive easily there.

"Here was an almost useless plant that had tremendous economic and botanic importance," Iltis said. "A weed sitting out there on the mountainside, barely used by anyone, yet you could estimate its potential value to our corn crop in billions of dollars."[5]

According to one estimate, the savings in diesel fuel alone—because plants would no longer be harvested and sown each year—could reach $300 million a year. The annual savings in preparing the fields and sowing corn could theoretically amount to more than $4 billion. Actual savings would probably be lower, because plants would occasionally have to be replaced and some plowing would be required. But there would be other advantages, too. Perennial corn plants, for example, would curb the loss of topsoil caused by repeated plowing and harvesting.

Fifteen years after its discovery, *Zea diploperennis* has not yet had an important impact on commercial corn. The crossbreeding of perennial corn has proven more difficult than researchers had anticipated. Some botanists have argued that perennial corn plants could not compete with conventional corn plants. It is a simple biological argument: Annuals put all of their energy into producing fruit; perennials must divide their energies between producing fruit and producing the rhizomes that allow them to survive the winter. The likely result is that perennial corn plants would be less productive than annual corn plants.

Some botanists argue, however, that this analysis only holds in temperate zones, where large-scale, mechanized agriculture is already established. Perennial corn might find its most advantageous use among small farmers in the tropics. The aim there is not to seek maximum yield from each acre but to produce as much as possible with the least input.[6] The savings in labor could be crucial for farmers earning a subsistence living.

Even if *Zea diploperennis* does not give rise to perennial corn, it is still fair to say that it will be worth billions. Shortly after it was discovered, researchers found that it had developed an amazing repertoire of defenses against diseases. It is completely immune to four diseases of corn, including the most important corn disease of South America and Africa.

For three of those diseases, no other source of immunity is known. One of the three is maize chlorotic dwarf virus, one of the most serious viral diseases of corn in the United States. *Zea diploperennis* is also partly immune to two other corn diseases. That is, they can infect it, but the infection does not do much damage.

The genes responsible for immunity in *Zea diploperennis* could have a dramatic economic impact if they were bred into cultivated corn varieties. If genes from *Zea diploperennis* were able to rescue only 1 percent of the U.S. corn crop from disease, the savings would be more than $250 million a year, researchers have estimated. Even if perennial corn never becomes a reality, the potential value of *Zea diploperennis* is beyond dispute.[7]

"It is not going too far to assert that those few thousand stalks of wild corn growing in a Mexican forest could ultimately rival, in dollar terms, the few hundred tree seedlings that were smuggled out of Brazil one century ago and thereafter used to establish the $4 billion rubber-growing industry in Southeast Asia," said the conservationist Norman Myers.[8]

Although Iltis's story has a happy ending, it raises disturbing questions. The location that Iltis and Guzmán stumbled onto was one of only three places in the world where *Zea diploperennis* grows wild. Iltis estimates that there are "perhaps one or two thousand plants of it in all the world." If Iltis and Guzmán had waited a month, a week, or even a few days before struggling up the steep slopes that led them to *Zea diploperennis,* they might never have found it. A single farmer raising a few head of cattle or clearing a field for maize could have unwittingly destroyed it. If they had not found that first population, they might not have continued their search. *Zea diploperennis* might never have been found.

How could such a valuable resource have come so close to disappearing? Have other similarly valuable plants become extinct without ever being seen by botanists? Are there still others out there that are near extinction right now?

While there is no way of knowing what has already been lost, it is likely that there are many other valuable plants waiting to be discovered. How close they are to extinction can only be guessed. But given the accelerating destruction of natural habitats around the globe, it is almost certain that many valuable food plants are disappearing. The consequences of this loss of agricultural resources are potentially enormous. Not only does it represent billions of dollars in losses to farmers; it could mean the difference between survival and starvation for millions of people around the world.

MENDEL AND THE TRIUMPH OF AMERICAN AGRICULTURE

Gene hunters like Iltis and Guzmán have done as much as any single group to confound Malthus's predictions that the world's expanding demand for food would outstrip the supply. One of the first applications of the rediscovery of Mendel's laws of genetics, around 1900, came in plant breeding. Biologists saw that they could improve crop attributes by judicious crossbreeding. They went to work, touching off a revolution in agriculture.

The revolution had enormous consequences, especially for American farmers. With less than 7 percent of the world's land and less than 5 percent of its population, the United States, in 1988, produced 11 percent of the world's crops, worth $73.5 billion. American farmers grow 50 percent of the world's soybeans and 40 percent of its corn. They lead the world in agricultural exports. In 1989, U.S. farm exports made up 12 percent of the world's trade in agricultural goods, more than twice that of second-place France.

Each of the 2.1 million farmers in the United States produces enough food and fiber for 128 people. An hour of farm labor now produces eight times what it did after World War II. American farmers are the most productive farmers in the world.[9] The principle problem with American agriculture often seems to be how to get rid of the surpluses.

Some of the increases in productivity are due to increasing reliance on fertilizers and farm machinery. In 1989, farmers' purchases of fertilizer and other chemicals were nearly ten times what they were at the close of World War II. Machinery costs were almost 50 percent higher.[10] But that does not explain all of the growth in productivity. There was a hidden component, too. American farmers can trace much of their success to one of the greatest and least appreciated advances of the twentieth century: the advent of scientific crop breeding.

The revolution set off by Mendel's research produced the most significant change in agriculture since it arose about ten thousand years ago. In every century until this one, "agricultural yields were low and unreliable," said Donald L. Plucknett, a scientific adviser to the Consultative Group on International Agricultural Research (CGIAR), an organization allied with the World Bank.[11]

"Agriculture was an uncertain business," Plucknett said. "Farmers had few options to change or alter their production system." Progress came partly from a primitive kind of crop improvement. Each year, at harvesttime, farmers would select the following year's seeds from the sturdiest and most productive plants in their fields. That was no guarantee of a better harvest the following year. But, over a period of years, it helped.

Crops would become a little tougher, a little more resistant to pests and diseases. Their yield crept slowly upward. The benefits were compounded as farmers bestowed their best seeds on their children, who would continue the process. As generations came and went and the process continued, each farmer eventually held seeds uniquely suited to his farm. Traditional crop varieties developed in this way are called *landraces,* and over the centuries, they accumulated by the tens of thousands.

Although landraces have been replaced by commercial varieties in industrial nations, they are still grown on small farms throughout the developing world. But the gains they provided over the years were modest. "Agricultural progress was slow and halting," said Plucknett. "Crop failures and 'bad years' were frequent . . . famine was a constant companion of human societies."

With the advent of the twentieth century, a new kind of crop improvement came along. Traditional methods of crop selection were supplanted with crop breeding based on the emerging science of genetics. Landraces arose from intelligent selection of seeds produced by chance crossbreeding in farmers' fields. Once researchers understood how genetic traits were passed on, they were able to select crops with desirable characteristics and crossbreed them in a way that would enhance those characteristics.

When two crop varieties were crossed, some of the offspring would inherit desirable characteristics from both parents. Those individual plants would be bred again with one or the other of the parent varieties, or with other varieties, to boost the number of desirable traits and eliminate undesirable ones. With repeated crosses, plant breeders could, in a few years, produce varieties that would never have arisen under the old landrace system. Those scientifically produced varieties were the foundation for the giant increases in productivity that have characterized the American agricultural miracle.

The breakthrough came in the garden of the nineteenth-century Austrian monk Gregor Mendel. Mendel devoted his life to the study of the inheritance of genetic traits in pea plants.

A student of science and mathematics, he entered a monastery in Brünn, where Iltis was born, in 1843, mainly to be free to pursue his scientific interests. There Mendel established an experimental garden. Over a ten-year period, he planted thirty thousand pea plants in that garden. He made crosses between peas with different characteristics and noted the results. He would fertilize a tall-stemmed plant with pollen from a short-stemmed one. And he would note the height of the offspring. Or he would cross plants that had wrinkled seeds with plants that had smooth ones. With each cross, Mendel would record the results.

As he analyzed the data that began to mount in his notebooks, patterns began to emerge. Mendel developed hypotheses to explain those patterns, and he devised new crosses to test the hypotheses. Working alone in the monastery in Brünn, Mendel derived all of the fundamental laws of inheritance. He single-handedly established the basis for the modern science of genetics. Mendel's laws apply not only to garden peas but to all sexually reproducing organisms, including human beings.

Mendel announced his laws to the Natural Sciences Society of Brünn in 1865. He published his findings the next year in the society's *Proceedings.* One of Mendel's central findings was that each genetic trait is governed by two "elements." The two elements could combine in different ways to produce different outcomes in an individual plant. Mendel further showed that each genetic trait was independent of every other trait. Elements that controlled a plant's height, for example, combined in ways that were completely unrelated to the combination of elements determining seed appearance.

Mendel's laws are a cornerstone of plant breeding and of all biology. But their significance initially eluded other biologists. His report in the *Proceedings* was ignored. The significance of his work was not apparent. He died without ever receiving the acclaim he was due.

Mendel's work dropped out of sight until 1900. Researchers in Germany, Austria, and Holland independently rediscovered his findings and recognized their significance immediately. Thirty-five years after he made his revolutionary discoveries, the times had caught up with Gregor Mendel. His place in the history of science was secured.

It was only a few years until the impact of this new science began to be felt in agriculture. Mendel made genetics a science by introducing the notion of experimentation, the idea of making deliberate crosses to try to elicit particular results. Mendel replaced chance crosses with planned crosses from which he could make scientific deductions.

His aim was to devise a theory to explain the passage of inherited traits across generations. He was not especially interested in crop improvement. But those who were quickly realized the value of Mendel's methods. In 1902, William J. Spillman of Washington State College replicated many of Mendel's findings in experiments with winter wheat, an important commercial crop. With those experiments, agriculture turned a corner. It was clear that scientific crossbreeding could greatly accelerate the process of crop improvement.[12] Instead of waiting for the appearance of desirable traits in the field, scientists could now produce tailor-made crops. The process of crop improvement lurched into overdrive.

The payoff has been more substantial than anyone could have expected. Between 1930 and 1980 yields of U.S. crops soared to almost unbelievable levels. The U.S. corn yield increased by 333 percent, from twenty-one bushels per acre to ninety-one bushels per acre. Wheat yield increased by 136 percent per acre. The yield of potatoes nearly quadrupled. Tomatoes surpassed that. In 1930, an acre planted with canning tomatoes yielded a harvest of four tons. Fifty years later, an acre of canning tomatoes produced twenty-four tons, an increase of 500 percent.[13]

Experts estimate that roughly half of the increase in each crop was due to the development and the widespread adoption of improved crop varieties. (The rest was due to fertilizers, pesticides, mechanization, and so on.)

To continue that increase, plant breeders wage a seesaw battle with pests and diseases. Crop varieties are replaced every five to ten years; the commercial lifetime averages about seven years. By that time, microbes and pests have adapted to a new variety and have begun to overwhelm it. Some varieties are overcome the first year they are used.[14] So breeders are constantly churning their resources, always working on the next variety so that it is ready when needed. That depends on a steady supply of new crop germplasm.

The Agriculture Department estimates that these improvements in crop varieties are worth $1 billion a year.[15] That is the payoff of a half century of skillful manipulation of the agricultural gene pool. And it is only the beginning.

The gains made so far, as impressive as they are, only hint at what might be possible in the future. Breeders have not yet come close to the maximum crop yields that are theoretically possible. Under controlled laboratory conditions, many crops have produced far higher yields than what has been achieved by American farmers in the field. The conclusion is that continued crop improvement still has a critical role to play in American agriculture.

In 1989, for example, American farmers produced an average of 116 bushels of corn per acre.[16] The world record for corn production is four hundred bushels per acre. But the theoretical maximum for corn production is even higher. Calculations suggest that with optimal sunlight, carbon dioxide, water, and nutrients—and without pests and diseases—corn yield could reach a staggering eleven hundred bushels per acre, nearly ten times the current U.S. yield.

Similar gains may be possible with other crops. The world record for wheat is seven times the average American yield. For rice, it is 5.8 times what American farmers get. Again, theoretical yields are likely to be even higher than these world records. Investments in crop improvement are likely to yield benefits for many years to come.[17]

"We've got the land, we've got the water, and we've got the air," said Kenneth Frey, a botanist at Iowa State University in Ames. "There's no way you can tap those to be useful to humans without going through a green plant."

VAVILOV AND THE TROPICAL CONNECTION

Mendel's discovery was the crucial first step in producing plants that could more efficiently tap the sun's energy to convert land, water, and air into food. But something else was necessary to produce the agricultural revolution of the twentieth century. After Mendel, botanists knew how to breed; what they then needed was a source of raw materials: plants with genes and corresponding traits that could be bred into cultivated plants to improve them.

The inexact science of gene prospecting came from the work of the second giant in plant breeding, the Russian scientist Nikolai I. Vavilov, whose pioneering research in the first half of the twentieth century took what Charles Darwin had learned and applied it to agriculture. Vavilov developed a comprehensive theory on the origin of domesticated crops and their interrelationships. In practical terms, his work identified the places in the world where crop relatives with the most valuable genes were likely to be found.

Not content with theory, Vavilov went in search of those plants. He was the first great agricultural plant explorer, visiting fifty-two countries and showing the way for many who came after him, including Hugh Iltis and Rafael Guzmán. Vavilov's research and exploration combined with Mendel's work to give an even greater impetus to agricultural research than either could have alone.

"His ideas continue to influence profoundly plant breeders, geographers, anthropologists, and those interested in the origin of agriculture," said Garrison Wilkes, a botanist and plant explorer at the University of Massachusetts in Boston. "Vavilov did for the geography of crop plants what Mendel did for genetics."[18]

Unlike Mendel, Vavilov is virtually unknown to the general public. Yet the story of his life is one of the greatest tales of scientific discovery in the twentieth century. It is also one of the most tragic. Some of Vavilov's colleagues gave their lives to protect his research. Vavilov's passionate defense of his scientific integrity ultimately cost him his life, too.

Vavilov was born into a wealthy manufacturing family in Moscow in 1887.[19] His father, hoping his son would one day become his partner, sent Vavilov to business-training school rather than to a classical high school. Although he mastered English, French, German, and Italian, Vavilov did not learn Latin and so could not enroll in a university. He was sent instead to the Petrov Academy of Agricultural Sciences. By this time, it was clear that Vavilov's passion was science, not business. (His brother likewise failed to join the family business, becoming a physicist and later the president of the Soviet Academy of Sciences.)

It was the dawn of the era of modern genetics, and Vavilov soon became a part of it. In 1913, he received a grant to study in England, where he met, among others, Francis Darwin, Charles Darwin's son. He left England after the outbreak of World War I. Vavilov attempted the dangerous journey home and arrived safely.

His research materials—all of his notes and his collections of specimens—were sent by a cargo ship. The ship hit a German mine. All of Vavilov's work was lost.

Vavilov had suffered eye damage while doing chemistry experiments as a child, so he was not inducted into the army. But the army soon saw a need for Vavilov's expertise. It asked him to help solve a medical mystery.

Russian soldiers who had liberated northern Iran began to be afflicted by a strange illness causing convulsions, hallucinations, and sometimes death. Doctors had traced the disease to bread baked from local flour, but they did not know why the bread was causing illness.

Before he left Moscow for Iran, Vavilov thought he knew the answer: The flour, he suspected, was contaminated by seeds of a poisonous weed. He was not terribly interested in the problem the authorities had given him, but he was eager to visit Iran. In his experimental plots at home, he had found a mildew-resistant variety of wheat that he thought might be important to Russian agriculture. The variety was identified only as "Persian." In 1916, he left for the Turkish front.

Working conditions at the front were dismal. Vavilov collapsed in the heat. He got sick. At one point, he was mistaken for a German spy. When his translator proved unreliable, he was forced to learn Farsi in two weeks. But Vavilov was right about the epidemic. The flour was indeed contaminated by a poisonous weed. The army, pleased with Vavilov's work, authorized him to undertake further botanical explorations.

Vavilov came home with a huge collection of wheat and rye seeds. That was the beginning of what would become the largest collection of seeds in the world, rivaled only by the NSSL in Fort Collins, Colorado. By 1941, Vavilov's seed bank contained more than 187,000 specimens. Vavilov continued his collecting expeditions, and his fame spread. In 1921, he moved to St. Petersburg to become head of the botanical institute, which later became the All Union Institute of Plant Industry.

From that post, Vavilov launched collecting expeditions around the world. Vavilov himself circled the Mediterranean, penetrated Afghanistan and Ethiopia, and traveled to Mexico and Central America, among many other remote locales. Before visiting each country, he would learn its language. At the end of his life he spoke twenty-two languages and could quote the Talmud and the Koran in the original.

The plant institute grew into a network of four hundred research laboratories with twenty thousand employees. Each expedition brought hundreds or thousands of seed samples back to St. Petersburg. Vavilov did not let the collection molder there. He distributed seeds to researchers across the Soviet Union for breeding experiments. His scientific contributions extended to archaeology, geography, and linguistics, as well as botany.

At the age of forty-one, Vavilov became the youngest scientist ever to be awarded full membership in the Soviet Academy of Sciences. His fame spread outside the Soviet Union, and he achieved the rare distinction of being named a foreign member of the British Royal Society.

One of Vavilov's central contributions to botany was his identification of what he called the centers of origin of cultivated plants. While Vavilov was roaming the world in search of seeds, he observed that in certain areas there would be an explosion of diversity in particular crops. He found a profusion of corn varieties in Mexico, for example. Wheat varieties seemed to be concentrated in the Middle East. These centers of diversity, Vavilov concluded, represented the centers of origin of the crops—the places where the crops had had the most time to evolve and diversify.

Each center of diversity was the origin of a dozen or more crops, by Vavilov's reckoning. China, for example, was the center of origin of soybeans, leaf mustard, apricots, peaches, oranges, sesame, tea, and several other crops. Other centers included India, where rice came from, and the Andean region of South America, where tomatoes and potatoes originated. (That is why Hugh Iltis went to Peru to look for wild potatoes and tomatoes.) There were eight centers of diversity on Vavilov's botanical map of the world.

In each center he found many crop varieties with a corresponding abundance of genetic traits useful for disease resistance and boosting crop yields. Some were cultivated varieties; others were the wild relatives of cultivated crops. Both were critical resources in plant breeding, Vavilov recognized, because both contained genetic traits that had not yet been incorporated into cultivated varieties. Those traits could boost disease resistance, increase yield, improve the quality of grains, and even help breeders tailor crops to local growing conditions.

"One would not be exaggerating in describing Vavilov's discoveries as a genetic gold mine," said the plant breeder Jack G. Hawkes.[20] In the 1930s, the United States, England, Germany, and Sweden began tracing Vavilov's footsteps, establishing their own collections. "Plant breeders had not previously had the remotest idea that such genetic richness existed," said Hawkes.[21]

Vavilov's seed bank, part of what is now known as the N. I. Vavilov All-Union Institute of Plant Industry, endures as one of his crowning achievements. It contains some 380,000 specimens brought back from more than 180 places around the world. Little known outside the circumscribed world of plant breeders and botanists, it is one of Russia's greatest natural resources. In a country heavily dependent on agriculture, the seed bank is critically important to the country's national security. Yet the collection almost did not survive the siege of Leningrad during World War II.

The crisis came during the unusually harsh winter of 1941–42. Hitler had blockaded Leningrad (St. Petersburg). He was unable to capture the city, but he continued shelling in a siege that lasted 880 days. As the battle wore on, little food reached the city. The population teetered on the brink of starvation.

The story of what happened next at the plant institute was told in a 1993 article by two of its scientists, Sergei M. Alexanyan, head of the institute's foreign relations department, and V. I. Krivchenko, a former director of the institute. When the siege began, according to the two scientists, technicians from the plant institute worked feverishly to duplicate the most important specimens in the collection, fearing that it might be destroyed.

Duplicates of many of the institute's potato specimens had been planted at an experiment station in Pavlovsk, twenty-five miles southeast of Leningrad. Pavlovsk was under fire, and it seemed likely to fall to the enemy. Despite the danger, workers rushed into the field, harvested the not-yet-ripe potatoes, and with the help of the Red Army brought them to Leningrad. The job was finished just days before Hitler's troops took Pavlovsk.

As the winter deepened, conditions at the plant institute became intolerable. "The building was unheated, as there was neither firewood nor coal," Alexanyan and Krivchenko wrote. "Because of unrelenting firing on the city's center, the building's windows were broken and had to be boarded up. The institute was cold, damp and dark."

The potato collection was stored in the institute's basement until March 1942. Workers frantically tried to keep the potatoes from freezing. "They burned everything to get heat: boxes, paper, cardboard and debris from destroyed buildings," the scientists wrote. The workers guarding the collection were "numb with cold and emaciated from hunger." The institute was spared from the bombs raining on Leningrad because it was located on St. Isaac's Square, near the Astoria Hotel. Hitler had ordered his troops to preserve the old hotel. He planned to hold his victory celebration there.

Meanwhile, as the siege continued, word spread among the desperately hungry people of Leningrad that the institute was bursting with potatoes, rice, and other edibles. Guards were posted. An emergency evacuation plan was devised to remove the collection, if necessary. Security inside the seed bank was tightened, too. The collection was housed in sixteen separate rooms. No one was allowed to remain alone in a room with the seeds.

Tens of thousands of people died of starvation in Leningrad that winter. Among them were at least nine scientists and workers at the plant institute. Surrounded by mountains of food, they chose to starve to death rather than eat the seeds in Vavilov's collection. Alexander Stchukin, a specialist in peanuts, died at his writing table. Dmitri Ivanov, a rice expert, perished, too, while preserving several thousand packets of rice. "As they slowly starved, they refused to eat from any of their collection containers of rice, peas, corn and wheat," Alexanyan and Krivchenko wrote. "They chose torment and death in order to preserve Vavilov's gene bank."

Vavilov himself was not there to defend his collection. He was in prison, the victim of a political attack on Soviet biological research.

Several years earlier, when Vavilov's scientific career was at its peak, the notorious Soviet biologist Trofim D. Lysenko launched a vicious attack against him. Lysenko rose to prominence preaching the pseudoscientific idea that characteristics acquired by an individual during his or her lifetime could be passed on to offspring. Stalin embraced the notion because it seemed more congenial to his ideas about the betterment of the Soviet state. In the battle between Lysenko and Vavilov, Stalin allied with Lysenko. Vavilov fought back. "They can send us to the fire, burn us, but we won't recant our principles," he said. Lysenko succeeded in having Vavilov removed as president of the Academy of Agricultural Sciences, which Vavilov had founded. In 1938, Lysenko replaced him.

On August 6, 1940, Vavilov was arrested while on a collecting expedition in Ukraine. He was interrogated for eleven months, charged with high treason and espionage, and sentenced to death. The geneticist who laid the foundation for the spectacular crop improvements of the twentieth century died of malnutrition in a Saratov prison on January 26, 1943.

Vavilov's seed collection remains one of the world's great crop collections. For plant breeders, the collapse of the Soviet Union has created an opportunity to establish seed exchanges with the Vavilov Institute. But it also poses a new threat to the collection, which is now facing its greatest crisis since World War II.

Russia's economic collapse threatens the government's ability to continue to support the seed bank. Seed banks, unlike museums and art collections, do not merely grow dusty if ignored. The collections at the Vavilov Institute are alive. Without proper care, the seeds will die, and the collection will be lost. That care includes maintenance of proper humidity and temperature and periodic replanting of the aging seeds in order to replace them with larger supplies of fresh seeds for distribution and experimentation.

Six of the institute's experiment stations, where seed regeneration is performed, were in different countries after the breakup of the Soviet Union. The five new governments that now have jurisdiction over the experiment stations—Ukraine, Turkmenistan, Kazakhstan, Uzbekistan, and Georgia—told the Vavilov Institute that they do not have the money to continue regenerating and evaluating the institute's collections. Instead, the governments have told the experiment stations to pursue research aimed at improving local farming. In Russia, meanwhile, the government has refused to spend its money on experiment stations located abroad.[22] Twelve of Vavilov's experiment stations are in Russia, but certain crops require climates found only outside the country.

The Sukhumi station in Georgia was destroyed in 1993 during Georgia's civil unrest, but at least some of its seeds were rescued. Alexey Fogel, an eighty-three-year-old botanist who had toiled at the station for fifty years, escaped through the Caucasus Mountains with his son, two other botanists, and 226 samples of subtropical fruit plants, including the station's collection of lemon seeds.[23]

The institute itself is also suffering. Beset by a shortage of hard currency, it canceled subscriptions to fifteen foreign scientific journals. Laboratory equipment is idled by the lack of cash to buy spare parts. Seed-collecting expeditions have been canceled. The institute is also having trouble attracting young scientists and technicians. "Commercial firms entice promising young researchers with incomparably higher wages, prospects of training abroad and better working conditions," said the institute's director, Viktor A. Dragavtsev.[24] An entry-level technician makes about $10.50 per month. The institute also sorely needs to establish a computer database describing the collection, Dragavtsev said. With the difficulties it is now encountering in regenerating seeds, it needs more refrigerated storage facilities so that the seeds can be kept longer before needing replanting.

The institute has managed to establish four new experiment stations inside Russia, and it has begun several new research projects. The U.S. government has promised the institute $1.5 million for personal computers, refrigerators, and support for the experiment stations outside Russia. American botanists are eager to explore the collection, which they believe contains samples that are not part of any other collection, including that of the United States.

If those seeds can be duplicated in the NSSL, they should be secure. The U.S. seed bank in Fort Collins, Colorado, seems to have everything the Vavilov Institute lacks. The American seed bank is not likely to be the target of a military attack. The United States is not poised to dissolve into separate republics, moving experiment stations outside its borders. Despite the inherent limitations of seed banks, the extensive U.S. collection ought to ensure that a substantial portion of the world's agricultural germplasm is in good hands, permanently protected and readily available to researchers and breeders. The reality, however, is quite different.

The NSSL is not nearly as secure as it might seem. Years of chronic underfunding, combined with little understanding of its importance even within the Agriculture Department, have raised serious questions about the condition of the nation's largest and most important seed bank.

Chapter 2

Seed Banks and Seed Morgues

In March 1986, the telephone rang in the office of James A. Webster, an entomologist working for the federal government's Agricultural Research Service in Stillwater, Oklahoma. Pat Morrison, in Lubbock, Texas, was calling to sound an alarm. Morrison, also an entomologist, had been inspecting wheat in Muleshoe, Texas, when he spotted a kind of aphid he had not seen before. He could not be sure what it was, he told Webster, but he was worried.

The appearance of a new pest always causes concern. But Morrison had a hunch about this one. If he was right, the aphid's appearance was an ominous sign for wheat farmers across the United States. Webster and his colleagues in Oklahoma City were worried about precisely the same thing. "We suspected it was the Russian wheat aphid," Webster recalled. "So we asked him to overnight some samples to us, preserved in alcohol." The Russian wheat aphid was known to be a serious threat to wheat, barley, and other crops. At the time of Morrison's call, researchers had not yet documented its presence in the United States.

Webster could not risk having Morrison ship live samples to

Oklahoma. That is why he asked that the samples be preserved in alcohol. The last thing Webster wanted to do was to accidentally contribute to the aphid's spread. However remote the chance that live aphids might escape from Webster's laboratory, it was a chance Webster did not want to take.

Preserved samples were also rushed to the Agricultural Research Service's headquarters in Beltsville, Maryland. There Manya Stoetzel, an aphid specialist, did a quick identification. She called from Beltsville several days later to confirm both Webster's and Morrison's fears: Morrison's aphid represented the first documented infestation of the Russian wheat aphid in the United States. The infestation posed a threat not only to Texas's wheat but to the entire U.S. wheat crop.

As soon as Webster heard from Stoetzel, he was ready to act. Knowing that the aphid would become his business sooner or later, Webster had prepared for battle with this new enemy. The Russian wheat aphid, green, oval-shaped, and about the size of a large grain of sand, was identified by a Russian entomologist in 1900. It is native to southern Russia, the Mediterranean region, Iran, and Afghanistan. For decades after its discovery, it was largely ignored. That changed in 1978. The aphid turned up in South Africa, where it provided a stark demonstration of the damage it could do.

Webster had reviewed South Africa's experience with the Russian wheat aphid. In his office was a copy of what he called the "Russian wheat aphid bible": the proceedings of a 1984 meeting in South Africa intended to summarize everything then known about the pest. "This was information that wasn't in existence anywhere else in the world," Webster said. The problem was severe in South Africa. In an average year, South African farmers were dumping pesticide on 40 percent of the country's wheat acreage to fight the new pest. In bad years, the figure was climbing as high as 75 percent. The South Africans were saving much of their wheat, but at the cost of enormous expenditures for pesticides.

When Morrison called Stillwater, Webster was working on another aphid called the greenbug. The greenbug is still an important wheat pest, but the Russian wheat aphid represented a much greater threat. Webster immediately shifted his priorities. Funds were directed toward gearing up research on the Russian wheat aphid. Researchers and technicians were moved out of the greenbug program and into the new Russian wheat aphid research project. Additional scientists were hired to boost the size of the staff. Researchers were transferred to Stillwater from other Agricultural Research Service laboratories. As the researchers prepared their battle plan, reports of Russian wheat aphid infestation began to come in from areas north of Muleshoe. Webster and his colleagues could do nothing more than chart the aphid's spread. There was no way to stop it.

Entomologists had dreaded the coming of the Russian wheat aphid for six years. The aphid had been spotted in central Mexico in 1980. Researchers watched as the aphid slowly and steadily made its way north. Once the aphid reached the United States, its spread accelerated. One month after it was spotted in Muleshoe, it was seen in the southeast corner of Colorado. It moved four hundred miles north that summer. By September 1986, it had spread across Colorado.

By 1992, it had spread across 28 percent of the dryland winter-wheat crop in thirteen states, with enormous variations among the states. The aphid infested only 1 percent of the crop in Idaho, for example, but it was found in 94 percent of the dryland winter-wheat acreage in Washington.

In the 1988–89 winter-wheat growing season, the aphid was seen in more than half of the 34.4 million acres of winter wheat grown in the western states. Pesticides were sprayed on 2.2 million acres, at a cost of $21 million. Combining the costs of pesticides and yield reductions, the aphid caused losses of $54 million in 1987, $130 million in 1988, and $92 million in 1989.

The figures include damage to wheat and barley, the most seriously affected crops, along with damage to oats, rye, and triticale, a wheat-rye hybrid. Weather is a critical factor in the extent of the epidemic in any given year, so the costs of the aphid vary from one year to the next. During the period 1987–92, the losses caused by the Russian wheat aphid amounted to slightly more than $850 million.

When Webster began his Russian wheat aphid research program, the only way to protect American wheat from the aphid's tiny jaws was to drench the wheat with millions of dollars' worth of pesticides. Webster's job was to find a better way. Something was needed to protect the crops that would ultimately eliminate or reduce farmers' exclusive dependence on pesticides. Not only were pesticides of concern as possible environmental contaminants, they are also costly. Money spent on pesticides comes right out of a farmer's bottom line. Extensive reliance on pesticides can easily push a marginally profitable farm into bankruptcy.

A further problem with the pesticide solution for the Russian wheat aphid is that pesticides are not entirely effective. When the Russian wheat aphid infests a wheat plant, it injects a toxin into the leaves and sucks out the sap. The toxin stimulates wheat to produce the chemicals it would normally produce only when stressed by drought. The long, narrow green leaves fade to yellow or white. New leaves fail to unroll and flatten as they emerge. Infested wheat plants sport long tubes instead of what they are supposed to possess: long, narrow leaves resembling outsized blades of grass. The aphids seek shelter inside the tubes. The rolled-up leaves put the aphids out of reach of pesticide sprays, which settle on the outside of the leaves.

While pesticides provide a short-term solution to the aphid problem, they are too expensive to be used for very long. An entirely different approach will be required for long-term protection. Plant breeders will be required to develop new varieties of wheat and barley that carry natural resistance to the Russian wheat aphid. Improved crop varieties do not foul the environment and impose little financial burden on farmers. Searching for new crop varieties was one of the most urgent priorities of Webster and his colleagues. The battle against the Russian wheat aphid provides a stark example of the importance of seed banks. The question was: Do the U.S. seed banks contain wheat varieties resistant to the new pest? That is what Webster had to find out.

When he began his search, he was not looking for crop varieties that could immediately be substituted for commercial wheat varieties. Commercial varieties are carefully bred to display a laundry list of desirable traits. Commercial crops are expected, above all, to produce high yields. They must be of appropriate size and uniformity to allow mechanical harvesting. They must mature at the right time to match local growing conditions. Wheat varieties must be nonshattering—that is, the plants cannot release the wheat grains to fall to the ground, as wild wheat plants do.

The grain that is harvested must meet consumers' and food processors' demands for appearance, taste, cooking qualities, and so forth. And crops must be bred to be resistant to a host of pests, diseases, and harsh weather conditions.

What Webster and others began to do was to search for any crop varieties at all that were resistant to damage from the Russian wheat aphid. If such varieties could be found, breeders would go to work to isolate the genetic traits that make them resistant to the new pest and crossbreed those traits into commercial varieties already possessing all the other required characteristics.

The resulting varieties—including Russian wheat aphid resistance, along with everything else—could then be sold to farmers in place of current varieties to reduce or eliminate farmers' reliance on pesticides. The failure to develop resistant crop varieties would hurt farmers, and it could easily result in a significant drop in wheat harvests, raising prices and hurting consumers.

When Webster and his colleagues began their search for resistant wheat varieties, they tested various crop varieties kept at the laboratory. "We had some material on hand, some barley, some wheat, which had greenbug resistance," Webster said. "We thought maybe we'll get lucky and some of this material will also have the Russian wheat aphid resistance. Well, it doesn't work that way."

The search for resistant strains quickly expanded. "First we looked at all the adapted cultivars grown in the United States," Webster said. (*Cultivar* is shorthand for cultivated varieties.) "What you want to do first is look at the material that's grown in the area. There may be some varieties in the United States that already have resistance." Again, Webster and his colleagues were disappointed. "We did not see anything that looked that good," he said.

Evidently, a successful search for resistant crop varieties was going to require the widest possible search of the available gene pool. The researchers turned for help to the U.S. Department of Agriculture's National Plant Germplasm System. There was no guarantee that Webster would find any varieties resistant to the aphid. After he failed to find resistance in American wheat cultivars, he began combing the riches of the seed banks. Wheat and barley varieties are stored in the National Small Grains Collection, which was then in Beltsville but was moved to new facilities in Aberdeen, Idaho, in 1988. Examining the collection's catalog, he first chose cultivars suited for U.S. growing conditions, for they would require less intensive breeding to produce suitable resistant varieties. When U.S. cultivars failed to show adequate resistance to the Russian wheat aphid, he began to search through seed-bank samples collected in the Middle East.

The reason for the focus on Middle Eastern varieties was that the aphid was thought to have originated there. That meant that wheat varieties that survived in the Middle East could have been exposed to the pest long enough to have evolved a natural resistance to it. Because of the volume of seed samples in the small-grains collection, limiting testing to Middle Eastern varieties was critical if the search was to be completed in time to save farmers from bankruptcy. The search was further complicated by the difficulty of testing for aphid resistance.

Testing for aphid resistance is considerably more involved than merely analyzing seeds in the laboratory. Because researchers do not know precisely which genetic traits will make a plant resistant, they cannot devise quick tests to detect resistance. Each sample to be tested must be planted in the greenhouse and grown to the point where it can make a meal for the laboratory's colony of Russian wheat aphids. (Once the aphids had already infested Oklahoma, the concern over having live specimens at the laboratory eased.)

"We plant sixty samples in a greenhouse flat, a metal box with soil in it," said Webster. "That way we can look at a lot of different lines. The plants are artificially infested. We maintain a huge colony of Russian wheat aphids in our rearing facility, in another greenhouse." Infested leaves from the aphid colony are snipped off and laid down among the test plants.

"Then we'll initially evaluate those for damage, and we'll rate each clump fifteen days after infestation and twenty-five days after.

Incidentally, all the scores are submitted to Aberdeen so they can maintain that on their computer, whether each variety is resistant or susceptible. If we do find some resistance in those small little clumps, of course we'll have saved some seed, and we'll retest that."

By the spring of 1991, five years after the aphid appeared in Muleshoe, Webster and his team had completed the laborious and time-consuming screening of wheat and barley varieties in the National Small Grains Collection. Webster searched ten thousand of the forty-three thousand varieties of wheat and its relatives and twenty thousand of the twenty-six thousand varieties of barley and its relatives. (The entire collection holds more than 112,000 samples of wheat, barley, oats, rice, rye, a wild-wheat relative called *Aegilops,* and a wheat-rye hybrid called triticale.) "To make a long story short, we have identified about twenty-five or thirty lines of wheat and maybe twenty lines of barley that have some pretty good resistance," Webster said. "Most of them are from other parts of the world, like from Afghanistan, the Soviet Union."

At that point, Webster began to scale back the screening program slightly in favor of research to discover precisely why these resistant strains are better able to withstand the aphid's attack. The information on the resistant varieties was made available to breeders across the country. "I think all of us feel real good about what we've done. In the first place, we were able to detect or identify the insect, disseminate the information, and get something going almost within a week of the time we were called by Pat Morrison."

The seed banks had delivered. With the resistant strains in hand, research could proceed. But the search also demonstrated why maintaining the largest possible gene pool is so important. A search of thirty thousand varieties of wheat and barley had yielded only a few dozen resistant varieties. Without tens of thousands of samples to search through, Webster might not have found any resistant varieties.

Even so, it had taken five years of effort to take the first step toward defending wheat and barley—two of the nation's most important crops—against a single predator that appeared one day in a field in Texas. And the job is not complete.

The resistant varieties identified by Webster and his colleagues are not suitable for distribution to farmers. The varieties contain a wealth of characteristics that make them unsuitable for widespread use. Some are poorly adapted to American growing conditions. The quality and characteristics of the grain are not right. Some may also turn out to be vulnerable to the greenbug or to other pests and diseases.

What breeders must do is perform a complex series of crosses between the resistant varieties and commercial cultivars. The aim, after many generations, is to produce commercial cultivars that are like their commercial parents in every respect except one: They have been fitted with resistance to the Russian wheat aphid.

BREEDING A TOUGHER NEW WHEAT

James Quick of Colorado State University is one breeder trying to develop wheat varieties resistant to the Russian wheat aphid. His aim is to breed resistance into varieties already well suited to Colorado's growing conditions. His work in Colorado is typical of what is happening in other wheat-growing states plagued by the Russian wheat aphid.

The aphid's economic impact on Colorado has been considerable. Winter wheat is Colorado's leading cash crop. It contributes $350 million to the state's economy. From the beginning of the infestation in 1987 until July 1990, the Russian wheat aphid caused losses of $83.5 million in Colorado, according to Frank Peairs, an entomologist at Colorado State University.

Part of that figure is attributable to the dramatic increase in pesticide use. Colorado's winter wheat is planted in September and harvested around July 4. In the 1984–85 season, before the aphid arrived, ten thousand to twenty thousand acres of Colorado were sprayed for various insects, Peairs said. Two years later, when the aphid really took hold, Colorado farmers sprayed 15 million acres, a thousandfold increase. That situation cannot continue, Peairs said. "There are lots of reasons why spraying is unacceptable, but the most important is: The growers can't afford it. The spray takes two-thirds of their profit margin."

Over the short term, researchers are looking for improved

pesticides. "They have to have something to buy time with," said Peairs. Biological control—using other insects to combat the Russian wheat aphid—may become feasible by the end of the 1990s, Peairs said, but that, too, is only a stopgap solution. "From all I've seen and heard, the way we handle the Russian wheat aphid will not change until we get a resistant variety. That's the key."

Quick faces daunting problems in the development of aphid-resistant wheat. Most of Colorado's winter wheat is used for baking bread. In addition to aphid resistance, Colorado wheat varieties must possess good baking and milling characteristics. They must also be resistant to drought and tolerant of heat. Quick's job is to alter Colorado wheat varieties to include resistance to the Russian wheat aphid without losing any of these other important characteristics.

About ten wheat varieties are responsible for virtually all of Colorado's wheat. One variety, designated TAM 107, occupies about half of the state's wheat acreage. That particular variety also illustrates the importance of adapting crops to local conditions. TAM 107 is unusable in eastern Kansas, where it falls prey to disease. But it is the leading wheat in western Kansas and Colorado, where cooler, drier conditions discourage the disease that infests TAM 107 in eastern Kansas.

In 1987, Quick began searching for wheat varieties with resistance to the Russian wheat aphid. After six months, nothing especially promising had turned up. Meanwhile, however, Quick was collaborating with Agriculture Department researchers on a separate project. They had planted a collection of wheat varieties in the field to evaluate their drought resistance. He noticed that one of the varieties, which had come from the National Small Grains Collection, was relatively free of signs of aphid infestation. "That was one of many that didn't have aphids. It was a suspect," he said. Further investigation revealed that this particular variety was more resistant to the Russian wheat aphid than any of the other varieties Quick had examined. The variety, designated T-57, originated on the Russian steppes.

Quick crossed it with a Colorado wheat variety. He crossed the progeny with the Colorado variety again, in a technique called *topcrossing*. The aim of the topcrossing was to breed more and more of the Colorado parent's characteristics into each succeeding generation while breeding out all the characteristics of T-57 except one: its aphid resistance.

T-57 has several qualities that make it unsuitable for commercial use. It is too tall, its stem is weak, the grain is soft and white (instead of red and hard, as desired), and it takes too long to mature.[1] With each generation, Quick selected offspring that came closest to having the characteristics he wanted: shorter, stronger stems, disease resistance, hard red grain, and faster maturity. He also made sure that each plant he selected for further breeding was resistant to the aphids in his experimental colony. The resistant plants most closely resembling the Colorado parent would then be selected for further breeding, and the process would be repeated.

With each generation, more of the appropriate commercial qualities from the Colorado variety appeared. The problem is that breeding takes time. Each generation requires the planting of seeds, special treatment to induce flowering, pollination, and maturation of plants. The plants with the right characteristics are selected, and seed is harvested for the next generation. With winter wheat, producing a generation takes at least twenty weeks.

Another complication is that breeding a single resistance gene into wheat may not be enough to repel the Russian wheat aphid. If resistance is built on one gene, then it takes only one mutation in the Russian wheat aphid to overcome that resistance.

The chance that a mutant will arise in a relatively short time after a resistant variety is introduced is distressingly high, as breeders have seen time and again. It is simply a matter of natural selection. Highly resistant wheat will be almost completely free of signs of aphid infestation. If one aphid out of countless millions acquires a chance mutation and overcomes the resistance, it will reproduce and spread wildly, since it has no competition from other aphids, which are unable to colonize the resistant variety. Because insects occur in astronomical numbers and reproduce quickly, the probability that such a mutant will arise spontaneously is high. When it does, the lack of competition assures that it will multiply.

To avoid that, Quick is looking for resistance that is strong enough to protect the crop but not so strong that it is likely to encourage the appearance of hardier aphids in a season or two. If tougher aphids appeared at that rate, Quick would not be able to breed new crop varieties quickly enough to stay ahead of them.

Quick's task as a breeder is to slow the aphids down, not stop them entirely. One way to combat this rapid mutation of aphids into superaphids able to overcome resistance is to load crops with multiple-resistance genes.

Resistance comes in several forms. One form is marked by varieties that pests, for whatever reason, prefer not to infest. A second kind results from developing crops that inhibit reproduction of the pests. A third form makes plants tolerant of pests—that is, the plants can withstand infestation without suffering. Quick's T-57 variety carries the third form of resistance: tolerance. Test samples in Quick's greenhouse quickly become covered with aphids, but they remain healthy.

That is the preferred form of resistance, because it is less likely to lead to a mutation. "It allows the insect to be relatively happy on the plant" without damaging the plant, Quick explained.

Since the discovery of T-57, Quick has found several other wheat varieties with resistance: one from Bulgaria and three from the Soviet Union and Iran. Using those varieties and others, Quick ultimately hopes to incorporate several forms of resistance into commercially useful wheat for Colorado. Of course, each new breeding strategy again requires years of crosses and making the right selections from the resulting plants.

In July 1994, Quick's painstaking work resulted in the harvest of seeds of a wheat variety designated E5W. Quick's efforts had been successful. It had taken seven years. "That's fast," he said. "Normally it takes ten."

E5W was one of thousands of progeny selected from the T-57 breeding program. "This is the only one that had all the requirements we need," Quick said, including resistance to the Russian wheat aphid. E5W resembles a cross between the widely grown wheat variety TAM 107 and a Colorado variety called *Yuma,* Quick said. And it includes the aphid resistance from T-57.

E5W, which Quick hoped to make available to farmers in time

for planting in September 1994, is not perfect. "It's important to point out that it's not quite as good in yield as the best variety out there," Quick said. "It's about 5 percent below. So a farmer is going to have to decide how bad his Russian aphid infestation may be" and then make a decision about spraying. The question is: Will the savings in pesticide purchases outweigh the costs associated with the 5 percent drop in yield?

"One spray costs ten to twelve dollars per acre," Quick said. "A quarter of the acreage was sprayed this year in Colorado. So twenty-five percent of the growers might plant it." In fact, Quick expects about 10 percent of farmers to use it the first year, with the number rising after that as better varieties come along. "We don't have it perfected to the level we would like it. But we will have lines two or three years from now that will be as good as what's grown." Farmers will then have resistant varieties with yields equal to what they are growing now.

Once that happens, however, Quick's work will not be over. Even the most effective forms of genetic resistance are eventually overcome by pests. Mutations inevitably lead to new "biotypes," or varieties, that are able to surmount the resistance that has been bred into crops. In the United States, the average lifetime for a variety of wheat, corn, oats, or soybeans is five to nine years. Even if the varieties continue to resist pests, they are usually replaced by higher-yielding varieties.[2]

Quick was lucky. He found a resistant variety more quickly than he expected to. The resistance came from a single gene rather than a combination of genes, making it easier to breed the resistance into commercial varieties. Breeders in other wheat-growing areas might not be so lucky. Nevertheless, without the immense collections in the federal government's seed banks, wheat growers in infested areas would be stuck with the use of costly, dangerous pesticides indefinitely.

THE NATIONAL PLANT GERMPLASM SYSTEM

In the shadow of the Rocky Mountains, on the campus of Colorado State University, lies one of the nation's most valuable natural treasures. It is not in any tourist guidebook. Nothing on nearby Interstate 25 alerts motorists to its presence. Students on the campus often return a puzzled look when asked about it. Indeed, there is not much to see. Signs identify it as the National Seed Storage Laboratory (NSSL). Its walls may contain more biological wealth per square foot than anyplace else on earth.

The laboratory houses 285,845 samples of seeds of modern, traditional, and wild varieties of the nation's major crops. The seed samples are packed in what look like paper-and-foil lunch bags, folded closed. The bags are stored on long rows of steel shelving in dark, refrigerated vaults. Most are kept at temperatures ranging from those of a kitchen refrigerator to about four degrees below zero. Other seeds and cuttings are packed into small tubes and lowered into a bath of liquid nitrogen for long-term storage at 321 degrees below zero.

The NSSL is as much an archive as it is a laboratory. It is sometimes described as the plant breeders' Library of Congress, offering an almost limitless selection of genetic traits for crop improvement. Because of its incalculable value to American agriculture, it is also sometimes called the plant breeders' Fort Knox.

The NSSL is the jewel in a nationwide network of seed banks known as the National Plant Germplasm System. Virtually unknown to anyone outside the small fraternity of plant breeders, these seed banks contain approximately 557,000 samples of seeds, plant cuttings, tubers, and roots.[3] It is the single largest seed collection in the world.

The collection is a storehouse of the accumulated wisdom of hundreds of years of crop domestication and improvement by farmers around the world. Its samples also incorporate the results of nearly a century of breeding by scientifically trained plant geneticists. The collection represents evolution frozen in time. Each sample is a living biological snapshot, preserving the evolved traits of a particular plant in a particular place at a particular time.

More than eighty-seven hundred plant species are represented, including not only the major crops but also beans, lettuce, sweet potatoes, peanuts, sweet clover, safflower, okra, gourds, beets, carrots, melons, and thousands of other important commodities.[4] Fruits and nuts are often stored as clippings rather than seeds, because in those species seeds don't reliably transmit all of the genetic characteristics of the trees from which they came.

The collection contains more than an assortment of cultivated crop varieties. It also includes many wild and weedy species related to the cultivars. The untamed species, unsuitable in many ways for cultivation, are often hardier and more resistant to disease and pests than cultivated crops, for the wild crops must survive without the careful nurturing of a human hand. Some of the seed banks also contain specialized scientific collections, including such things as seeds from plants with unusual genetic or chromosomal abnormalities.

Many of the seed samples are from plants that no longer exist in the wild. That makes the samples irreplaceable and priceless. The value of the U.S. seed-bank collection to agriculture is impossible to estimate. The discovery of a single seed sample with resistance to the Russian wheat aphid could be worth at least $100 million a year to farmers. That alone would justify the claim that the seed collection is one of the nation's most valuable natural resources. And that is only one of countless examples of how the germplasm collection has saved farmers millions of dollars.

The enormous size of the seed-bank collections might sound like overkill. No breeder, in a lifetime, can hope to explore completely the thousands of samples of even a single crop. Breeders insist, however, that the enormous diversity represented in the seed-bank collections is critical. In fact, they say it does not suffice. "It is clear that the more genetic diversity that can be available to the breeder, the wider range of choice he will have in selecting the appropriate kinds of diversity for his breeding programs," said Jack G. Hawkes. A breeder might finally use only a fraction of what is available, but having the choice is more important now than it ever was, Hawkes said. In the early decades of the twentieth century, breeders still had ample opportunity to devise new crop varieties from the landraces and varieties available in their own countries. "In the last 50 years or so, however, a much wider range of genetic diversity is required," Hawkes wrote. The demand has been spurred by a growing population and a finite supply of agricultural land. The pressure on breeders to boost crop yields is greater than ever.[5]

The mission of the National Plant Germplasm System is no less than to provide "the genetic diversity necessary to improve crop productivity and to reduce the genetic vulnerability in future food and agriculture development, not only in the United States but for the entire world," according to the Agriculture Department's description.

SEED BANKS IN CRISIS

Considering the importance of plant genetic resources, it would seem that guardianship of the nation's seed banks ought to be accorded the highest national scientific priority. A century ago that was true. Seed collecting was the Agriculture Department's highest priority. In 1878, it spent one-third of its budget on the collection of crop germplasm. The American farmers who received it "provided a testing ground and ultimately created the genetic base on which industrial capitalism could be founded," according to the historians Jack R. Kloppenburg, Jr., and Daniel Lee Kleinman.[6]

One hundred years later, the situation is quite the opposite. The seed-bank system is the tragic victim of overwhelming official neglect. A dedicated group of professional plant breeders and agricultural scientists is holding the system together, even expanding it slightly. But they have failed to attract anything more than marginal support for their enterprise. Experts agree that at present funding levels the integrity of the seed banks cannot be maintained.

Part of the problem lies in the nature of the Agriculture Department. In the words of a General Accounting Office report released on the eve of President Bill Clinton's inauguration, the department has become "a twentieth century dinosaur" that has scarcely changed since the 1930s. The Agriculture Department estimates that the germplasm in the seed banks has been responsible for crop improvements worth $1 billion annually. Yet the budget of the seed banks—$30 million—is too small to rate the status of a separate line in the department's budget. In Congress, where the budget is reviewed, that is far too small to be noticed. As Congressman George Brown, the former chairman of the House Science Committee, put it, "That's not even budget dust." The department's food assistance programs spend three times that much *every day*.

Why are the seed banks so undervalued? The answer has something to do with the peculiar nature of plant genetic resources. It is true that they are enormously valuable, and many are exceedingly rare. Unlike the gold in Fort Knox or the rare jewels in the Smithsonian, however, the treasure in the seed banks can be made available in unlimited quantities. Seeds are a renewable resource. And this biological treasure is given away free. Any bona fide researcher who wants seed samples gets them—at no charge.

These peculiarities have led to a dangerously casual attitude about the riches in the seed banks. Sometimes the consequences have been disastrous. A case in point occurred in the late 1960s, when seeds from about five thousand varieties of tropical corn were sent from Mexico to the NSSL in Fort Collins. The precious samples were the product of an ambitious effort to collect and preserve rare corn varieties from all over Latin America. The samples sent to Fort Collins duplicated a collection kept outside Mexico City at the International Center for the Improvement of Maize and Wheat, known by its Mexican acronym as CIMMYT (pronounced *SIM-it*).

After the seeds were dispatched to the United States, CIMMYT officials forgot about them. A few years later, budget woes at CIMMYT nearly forced a shutdown of its seed bank. Some of CIMMYT's samples were lost. But researchers there did not worry. The backup collection was safely stored in Fort Collins. Or so they thought. What CIMMYT officials didn't know was that the duplicate collection at the NSSL had never been entered into the permanent collection. The samples had disappeared. No one could explain what happened.

Prof. Major Goodman, a widely known and respected crop scientist at North Carolina State University, decided to investigate. Goodman is an authority on Latin American corn varieties, one of a handful of researchers who understood the value of the missing samples. Goodman was stunned by what he discovered. "It turned out that one day, in cleaning up Fort Collins, somebody found these samples and didn't know what to do with them," Goodman said later. The seeds were stored in small, old packages and were clearly not part of the regular collection.

The laboratory wrote to CIMMYT to inquire about the samples. CIMMYT officials apparently were confused about which samples the laboratory was referring to. CIMMYT officials told administrators in Fort Collins that CIMMYT's seed bank had duplicates of all the samples. CIMMYT asked that some of the seeds be returned. The rest were simply thrown away.

As it happened, the CIMMYT officials erred when they told the NSSL to discard most of the collection. "CIMMYT did not have what they said they had, and no one did any cross-checking," said Goodman. "A lot of the stuff that was thrown out was lost." According to Goodman, no one at Fort Collins made any effort to contact him or any other researchers with special knowledge of Latin American corn. "If I'd known it was happening, I could have stopped it," Goodman said ruefully. "So could a dozen other people. But no one ever asked." Researchers can only guess what valuable genetic traits the lost seeds might have carried. There will be no chance to find out.

Officials who were then at the seed-storage laboratory denied that the seeds were thrown away. But Goodman insisted that was what happened, and the laboratory staff could provide no alternative explanation.

But that was only the beginning of Goodman's indictment of the seed-storage laboratory. As he continued to examine the collection, he concluded that virtually all of its samples of tropical corn were in jeopardy. On that point, there is less disagreement. The present administrators of the Fort Collins collection admit that many of the seeds stored in the laboratory's refrigerated vaults were not replanted when they should have been. Many of those probably will not germinate now. They are dead.

"There might be five thousand to ten thousand tropical corn varieties," said Goodman. "I know that there are problems, because they haven't been grown." Some date back to the 1950s, many to the 1960s, and a few to the 1970s and 1980s. "All are in potential trouble," he said.

Samples from the 1950s have likely deteriorated to the point where only about 50 percent of the seeds in each sample will sprout, he said. The seeds in each sample are not genetically identical. Like a roomful of people, a seed sample is made up of individuals, each with its own set of genes. If 50 percent of the seeds in a sample have died, many of the genetic traits in the original sample have disappeared with them. If the seeds are not replanted and regenerated, genes will continue to disappear as the sample ages and continues to deteriorate. "Fifteen percent to thirty percent of corn collections are sufficiently low that I'm concerned about it," Goodman said. "And I'm afraid that twenty years from now we're not going to be doing any more than we're doing today."

For seed samples collected before 1950, the picture is even worse. George White of the Agricultural Research Service estimated that perhaps 33 percent of the samples collected in the 1940s survive. That figure drops off for earlier samples. Of seeds collected in the 1930s, only about 11 percent can still be found in U.S. collections.

Between 1929 and 1931, seed explorers brought four thousand samples of soybeans back from China. Two-thirds of the samples have been lost or thrown away. One thousand samples of sugarcane were collected from Papua New Guinea beginning in 1875. A century later, in 1975, only 204 remained.[7] R. Dean Plowman, the administrator of the Agricultural Research Service, estimates that 90 percent of the seed samples brought into this country before 1950 were lost "due to inadequate knowledge and lack of suitable storage facilities."[8]

SEED-BANK BEGINNINGS

The U.S. seed collections were not always subject to official neglect. They have a long and distinguished history. In the early days of the republic, when farming was the predominant occupation, some of the nation's most famous farmers took a great interest in seed collecting.

One of them was Benjamin Franklin. That "champion of good sense and experiment," as the historian Daniel J. Boorstin called him, was, like many of the founding fathers, a farmer. When, in the service of his country, he joined the diplomatic corps, he spent much of his time encouraging the introduction of new plants to the New World. His diplomatic sojourns abroad nearly always doubled as seed-collecting trips. Although he is better known for his work on the study of electricity, his contributions to agriculture helped transform the economy of the New World.

Franklin made one of his most important discoveries while serving in England as the agent for the colony of Pennsylvania. There he came across some interesting beans that were known as Chinese garavances. The beans, which could be made into a kind of cheese, fascinated him. He sent a few home with instructions that they be distributed to farmers willing to try planting them.

The Chinese garavance is now known as the soybean. The "cheese" that interested Franklin was tofu. The beans that Franklin sent home gave rise to a soybean harvest worth $10 billion each year to American farmers. Half of all the soybeans produced in the world are grown in the United States. Among Franklin's other contributions to American farmers were rhubarb, upland rice, and broomcorn. Not to mention *Poor Richard's Almanack*.[9]

Franklin was not the only prominent American of his day to introduce foreign seeds to the struggling new nation. George Washington, Thomas Jefferson, and other farmers and landholders made a point of collecting seeds during their travels abroad.

Jefferson was one of the most dedicated of the early plant explorers. He viewed his seed-collecting experiences as among his most valuable contributions to the United States. In a short, reflective piece written in 1800, Jefferson listed what he thought were his most important contributions to his country. The Declaration of Independence was one. His work to promote freedom of religion was another. But, in his opinion, the most important one was something else. "The greatest service which can be rendered to any country," Jefferson wrote, "is to add an useful plant to its culture."[10]

When Jefferson was minister to France in the 1780s and 1790s, he arranged for annual shipments of seeds to America from the Jardin des Plantes in Paris. He obtained rice varieties from China, Egypt, Palestine, and Africa.[11]

On one occasion he risked the death penalty to collect seeds for his country. In 1787, Jefferson went to Italy to look for samples of Piedmont rice. Parisians favored it over those varieties grown in the United States. The Piedmontese, seeking to preserve their monopoly, guarded it carefully. But Jefferson would have none of that. When he found a source of the rice, he immediately made plans to smuggle it out of Italy. He hired a mule driver to take some to France. As insurance, he slipped a few handfuls into his pockets. A short time later, a shipment of the rice reached South Carolina farmers. At the time, the export of Piedmont rice was punishable by death.[12]

By 1819, seed collecting had received the official endorsement of the U.S. government. Secretary of the Treasury William L. Crawford directed America's diplomats and naval officers to send home any potentially useful seeds they encountered during their service abroad.[13]

Crawford's order produced the desired results. Seeds arrived from all over the globe. They were distributed by the government to any farmer who asked for them. The enterprise grew. By 1862, the government was distributing more than 1 million seed samples to farmers each year. Abraham Lincoln established the Department of Agriculture that year, partly to continue and expand the seed exchange.[14]

Without the efforts of Franklin, Jefferson, and countless others who gathered seeds from abroad, U.S. farmers would not be able to feed themselves, let alone produce the huge surpluses that are now taken for granted. The early seed collectors understood America's relative disadvantage in terms of native crops available for agriculture.

As Vavilov later discovered, nearly all of the world's major crops come from centers of origin loosely clustered around the equator. If the world is divided into the "haves" and "have-nots" of crop resources, the United States is one of the have-nots. It depends almost entirely on crops that originated outside U.S. borders and had to be imported. Only corn, of the major crops, is native to North America. It was brought to the United States from Mexico.

More than three hundred years before Franklin and Jefferson began stowing seeds in their pockets, Columbus began the greatest seed swap in history. When he landed on American shores, he saw local people growing corn, beans, squash, and sweet potatoes—things he had never seen before. He brought with him seeds of wheat, chickpeas, melons, onions, radishes, salad greens, fruits, and sugarcane, which he introduced to the New World. He took seeds of the American crops home.

A meal of foods native to the United States would be a slim one. Blueberries would be on the menu. But with no wheat for the crust, there would be no blueberry pie or pancakes. A glass of cranberry juice would be included, but it would probably be undrinkable without sugar to sweeten it. The main course would consist of sunflower seeds, pecans, and a few other nuts. The vegetable would be the Jerusalem artichoke, a tuber. Raspberries and grapes round out the list. Forget the Thanksgiving Day staples; squash, sweet potatoes, and pumpkins came from Latin America.

Three crops provide 60 percent of the calories and 56 percent of the protein that people get from plants: wheat, rice, and corn. All belong to a single family of plants, the grass family. The grasses also include barley, sorghum, millet, oats, and rye. The grass family provides about 80 percent of the calories that humans consume. None of these grasses originated in the United States. (Nor did Kentucky bluegrass, despite its name. It came to the United States from Europe, and it may have originated in Asia.) The same is true of many other kinds of crops. Of the world's twenty most important food crops, not one is native to the United States.[15]

That is why the U.S. seed banks are so important to America's bountiful harvests. U.S. plant breeders looking for useful genetic traits in the major crops cannot search their backyards. Wild relatives and traditional varieties of wheat, corn, and all the other important crops do not grow in the United States. They survive only in the regions where the crops originated. American crops, as one writer put it, "are powered by foreign genes just as surely as our industry is powered by foreign oil."[16]

The early American colonists recognized that they had settled in a have-not region. That may have contributed to America's eventual agricultural success. The colonists were forced to bring their own seeds with them. A remarkable diversity of crops sprang up where sunflowers and a few fruits and nuts had been grown before.

Seed saving is a centuries-old tradition dating back to the origin of agriculture ten thousand years ago. Seeds were buried, stored in caves, or set aside in specially built huts to preserve them for the next growing season. Indigenous people devised a variety of clever schemes for protecting and preserving germplasm critical to their food production. The Kayapó Indians of Brazil maintained "gene banks" in the form of hillside gardens. These gardens would preserve representative samples of important crop varieties in the event of floods or other catastrophic damage to crops planted at lower elevations.[17]

The Tohono O'odham Indians of the American Southwest preserved seeds in clay vessels with narrow necks. Lids were glued onto the vessels with a resin from creosote bushes, and the pots were stored in caves. The seeds, protected from insects, rodents, and heat, could survive for a decade.[18] The first recorded plant-collecting expedition was launched by Queen Hatshepsut of Egypt in 1500 B.C. Bas-reliefs in the temple at Deir el Bahari, near Luxor, describe the queen's dispatch of a team to the land of Punt, now Somalia, in search of the incense tree.[19]

In the eighteenth and nineteenth centuries, colonial powers established botanical gardens at home and in their colonies for the collection, preservation, and improvement of medicinal and food plants. The exchange of plants often had a substantial and unpredictable impact on local economies around the world.

In 1706, for example, a single coffee plant from Java was taken to the Amsterdam Botanic Garden. In 1713, the burgomaster of Amsterdam sent progeny of the Java plant to Louis XIV. The plant was nurtured in the Jardin des Plantes in Paris. Its offspring were dispatched across the Atlantic to Martinique in 1720. Only a single plant survived the rigors of the journey, but it gave rise to a coffee industry in Martinique. Plants from Martinique, in turn, were taken to Jamaica, where they were used to develop a premium coffee variety called Blue Mountain, considered one of the finest coffees in the world. Offspring from the Java plant were also sent to Suriname and Malawi.[20]

The U.S. National Plant Germplasm System is an outgrowth of the Agriculture Department's Section of Seed and Plant Introduction, established in 1898. The agency was established in response to demands by farmers for more and better crop varieties. Cattle ranchers in the Northwest were being displaced by farmers, who needed cold-tolerant crop varieties. In the Southwest, farmers were looking for drought-resistant crops.

As one of its first projects, the Section of Seed and Plant Introduction established a system of designating plants with plant-introduction numbers. Plant-introduction number one was a cabbage called Bronka that had been collected outside Moscow. The numbering system, still in use, has now cataloged more than five hundred thousand plant varieties.

The plant-introduction office coordinates the acquisition of germplasm, prepares the documentation for newly introduced varieties, and assures that plant-quarantine requirements are met for varieties introduced from overseas. Plants often arrive with a healthy cargo of troublesome stowaways, including weeds, insects, and disease-causing viruses and bacteria. Quarantine procedures are designed to eliminate these fellow travelers before foreign plants are made available for research and breeding programs.

In the early part of the twentieth century, American plant explorers were sent around the world on collecting expeditions, often enduring considerable hardships before returning with their booty. Frank Meyer was one of the first and one of the most daring American explorers. The Agriculture Department dispatched him to Europe, Russia, Tibet, and China three times in search of ornamental plants and crops, especially tough crops resistant to cold and drought, which often meant that Meyer had to endure cold and drought. His fortitude was legendary. He would rise before dawn, work all day in freezing rain, and sleep outdoors in subzero temperatures. Disease epidemics were common. Meyer and other explorers were frequently threatened by soldiers or thieves. As his biographer wrote, "He walked thousands of miles over lofty mountains and parched deserts, through snowstorms and dust storms, and into primeval forests never before seen by a white man. . . . He faced brigands in China. . . . He fought off an attack by three murderous ruffians in Khabarovsk and barely escaped being shot by soldiers in the Kansu Province of China."[21]

Unfortunately, in an age before refrigeration, many of the samples for which the early explorers risked their lives did not survive the journey home. But a few of Meyer's finds did survive return passage and have become quite valuable. A spinach variety he brought back from Manchuria was used to develop a disease-resistant spinach that rescued the spinach-canning industry in the United States. He retrieved a disease-resistant wild peach from northern China that has been used to provide rootstock for grafts of apricots, peaches, and plums. Siberian and Chinese elms collected by Meyer were used to plant seventeen thousand miles of windbreaks in the Great Plains during the 1930s. His contributions also include a popular Florida grass, from China, and a legume used to control erosion along interstate highways, from seeds found in the Soviet Union. In thirteen years of exploration, he introduced twenty-five hundred plants to the United States. His career ended tragically and mysteriously in 1918 during an expedition to China, when his body was pulled from the Yellow River. The circumstances of Meyer's death are not known, but his demise appears to have been the result of either suicide or murder.[22]

THE NATIONAL PLANT GERMPLASM SYSTEM

The seed-preservation network now known as the National Plant Germplasm System began with the passage in 1946 of legislation authorizing the establishment of regional plant-introduction stations. They have become the principal repositories for many of the nation's crop-seed collections.

The NSSL was completed in 1958. It is intended to serve as the backup archive for the regional stations, maintaining duplicates of everything stored at the stations in case some calamity should damage or wipe out the regional collections.

By the 1970s, the plant-introduction stations and the NSSL had evolved into a loose cooperative network of about three dozen germplasm collections stored in state and federal research facilities around the country. The network is administered by the states through their agricultural experiment stations and by the federal government through the Agriculture Department's Agricultural Research Service and its Cooperative State Research Service. Many of the collections serve as global resources. Germplasm is routinely exchanged with researchers, government agencies, and other seed banks around the world.

Most of the National Plant Germplasm System's seed samples—or accessions, as they are called—are distributed among seventeen federal and state facilities. The NSSL holds the lion's share of them, with 232,000 accessions. The NSSL's collection is not a so-called working collection. Its germplasm is rarely made available to breeders. Most of the material it contains is duplicated in working collections at other facilities. The situation is analogous to that of a large art museum. Some of the museum's collection might be circulating or on loan. Some is on display. But most is in long-term storage in rooms carefully controlled for temperature and humidity. Access is provided only to scholars by special arrangement. Similarly, the NSSL provides long-term secure storage but little access. In contrast, the samples in working collections, like museum artifacts on display, are widely available for study.

The next largest U.S. seed bank after the NSSL is the National Small Grains Collection in Aberdeen, Idaho. It houses more than 112,000 accessions of wheat, oats, rice, barley, rye, and two other less familiar plants: *Aegilops,* a wild-wheat relative, and triticale, a wheat-rye hybrid. "We have representations in the collection from every wheat-growing area in the world—Afghanistan, Turkey, Iran, Iraq, and Syria," says its curator, Harold E. Bockelman. Most are landraces collected by plant explorers during the last four or five decades.

About 135,000 accessions, or roughly one-third of the national collection, is stored at regional plant-introduction stations in Pullman, Washington; Ames, Iowa; Geneva, New York; and Griffin, Georgia.

Most of the germplasm system's holdings are seeds. Grains and other major crops are genetically constructed in such a way that their traits can be reliably passed on in seeds. This is not true of all crops. The desirable characteristics of most fruits, for example, develop in such a way that they are scrambled and changed during fertilization. The culprit is sex. As is the case with children, the offspring of fruits show some of the characteristics of each parent. They may also display a few surprising characteristics that seem to be all their own. Saving the seeds of fruits, therefore, provides no guarantee that the desirable traits of the parent variety will be preserved.

For these crops, cuttings, rootstocks, or living specimens must be maintained. A cutting from a fruit tree will give rise to another tree that is a clone, or genetically identical twin, of the first plant. Preserving living plants is quite different from saving seeds, and it is far more costly. To assure the survival of valuable clonal germplasm, the Agriculture Department has set up a nationwide network of clonal germplasm repositories at ten locations around the country.

The repositories hold twenty-seven thousand samples of a total of three thousand different plant species, including apples, citrus fruits, peaches, pomegranates, blueberries, pecans, hickory nuts, filberts, dates, and a variety of tropical fruits, many maintained in Miami, Puerto Rico, and Hawaii. The Agriculture Department's research station in Geneva, New York, is expanding its apple orchards to thirty acres with twenty-five hundred different varieties of apple trees. Some two thousand varieties of grapes are planted in the station's vineyards. It is the largest collection of grapes and apples in the world.

The Miami research station's holdings include nineteen hundred samples of sugarcane and related grasses. Some of the sugarcane varieties are descendants of samples collected by William Bligh in Tahiti in 1793. Bligh brought the cane back aboard the

Providence, the ship he captained after losing the *Bounty* in the mutiny led by Fletcher Christian.

A FAILURE TO EVALUATE

Unlike the treasures in the Library of Congress or Fort Knox, the treasure in seed banks is alive. Each seed is a tiny, breathing organism with a finite life span ranging from several years to several decades. A seed survives only as long as its internal energy reserve lasts. When it is depleted, the seed dies. No seed lives forever.

The maintenance of seed collections therefore requires that each seed sample be planted periodically to produce fresh seeds. The regenerated seeds can then be returned to the seed collection and distributed to other collections or to researchers. Periodic replanting of the seeds serves a second purpose, too. It allows seed-bank curators to evaluate a sample's potential value to breeders.

Seed samples are generally accompanied by what breeders call passport data, which should include when and where the sample was collected, along with field notes concerning its characteristics, how it interacted with other nearby plants, whether it appeared to be resistant to pests or disease, and so on. But the passport information does not tell breeders everything they need to know.

Critical questions about a crop plant, including its size, the time to maturity, its yield, hardiness, and insect and disease resistance, can be determined only by planting the seeds and examining and testing the resulting plants. The process is called evaluation. Examination of the seeds themselves will not do it. A seed's shape, size, and color say nothing about the characteristics of the plant it will produce.

Samples without adequate descriptive data are useless to breeders. Searching for useful traits in seeds that have not been properly evaluated would be like searching through a Library of Congress filled with books with no titles or descriptions. What you want may be there, but finding it is impossible.

To determine whether a particular crop variety is resistant to a troublesome bug, researchers use the crude technique that James

Quick used in his battle against the Russian wheat aphid. They grow plants and dump bugs on them. It is simple but tedious and time-consuming. Quick had one advantage: Russian wheat aphids can be easily raised in the laboratory. Some other pests cannot. Evaluation then requires that plants be grown in infested areas, where conditions cannot be controlled as well as in a greenhouse.

A further difficulty is that some plants require special growing conditions. Tropical corn plants, for example, cannot be grown at all in the United States, either in greenhouses or in the field. The problem is that corn has a curious sensitivity to the length of the day. Tropical corn is accustomed to ten to twelve hours of daylight each day, no more, no less. Planted in the United States, where daylight lasts hours longer during the summer, tropical corn will not flower, and therefore it will not produce new seeds. It will simply keep growing until it is fifteen to twenty feet tall. To regenerate and evaluate tropical-corn samples, U.S. seed-bank curators must make arrangements with breeders in tropical countries to plant the crops, evaluate them, and harvest the seeds. That adds to the cost of the process.

Regeneration and evaluation of seed samples are the first things seed-bank curators ought to do to make their collections useful to breeders. But that is only the beginning. The next task is what crop scientists call germplasm enhancement, or "prebreeding."

Most of the samples in the seed bank are not suitable for commercial cultivation. It can take two decades or more to separate useful traits from undesirable ones. Farmers whose crops are being ravaged cannot wait two decades for help. Enhancement, or prebreeding, is a process in which potentially useful genes are transferred into varieties more closely related to commercial crops. These interim varieties can be more easily crossed with the breeding lines used to produce commercial crops. That shortens the time it takes to get useful genes into commercial crops.

In both areas—evaluation and prebreeding—the U.S. seed-bank system has failed.

"A germplasm system which acquires accessions with the ability to evaluate and regenerate is a facade," said Major Goodman. "Evaluation, regeneration, and utilization are essential parts of a functioning germplasm system. Yet the entire emphasis . . . is based upon acquiring larger and larger numbers of accessions to be stored in so-called seed repositories." A more accurate name for them, Goodman said, is "seed morgues."

Huge seed banks like the NSSL seem to guarantee the survival of critical crop germplasm, but it is a false reassurance, Goodman said. He half seriously suggested that it might be better to get rid of the current system altogether. "With no germplasm system in place, perhaps a viable one would be demanded, rather than settling for the current masquerade."[23]

There are, for example, thirty thousand varieties of corn from Latin America. Goodman said he knew of only four scientists doing the work of regenerating and evaluating those samples. Each does about thirty samples a year, for a total of 120 samples regenerated and evaluated annually. At that rate, it would take 250 years to regenerate the entire collection. Samples may survive for decades, but certainly not for hundreds of years. If the regeneration program is not expanded, many of the corn samples might not survive much longer.

The effort to preserve the gene pool in the seed banks "is in full retreat," said Goodman. Without those genes, American agriculture could pay a terrible price. The crisis could occur suddenly, in the form of a voracious pest that suddenly overwhelms and destroys a major crop. Or the crisis could develop gradually as a succession of new pests evolves, each taking another small bite out of farmers' harvests. The cumulative effect of such pests, if plant breeders do not have the genetic raw material they need to fight them, could, over time, be even more devastating to agriculture.

The Agriculture Department does not have the resources it needs to regenerate and evaluate its collections. Prebreeding is even farther down the priority list. "Much of the germplasm in the [National Plant Germplasm System] is exotic or wild material, which is difficult to use in conventional breeding programs," said William W. Roath of the Agriculture Department's Plant Introduction Research Unit in Ames, Iowa. Prebreeding of this material "has been designated part of NPGS responsibility," but "few funds have been allocated for this purpose."[24]

The Agricultural Research Service has taken some steps to rectify the situation. But the administrators of the germplasm system have said that they have difficulty persuading even their own superiors that they need more money. Without the support of top Agricultural Research Service administrators, there is no chance of additional support from the Agriculture Department or Congress.

"The costs of not establishing a functioning germplasm system—rather than a façade of such a system—are far higher than the costs of an adequately run system," said Goodman. "Indeed, the return from a single major disease or insect resistance factor in any one major crop, such as wheat or corn or cotton, would easily pay for all agricultural research expenditures, both private and public, for the entire country for at least a decade."

Some successor to the Russian wheat aphid will one day prove resistant to everything that has been evaluated in the seed banks. Resistance to the aphid might be hidden somewhere in the unevaluated samples, or it might have disappeared in samples that were allowed to deteriorate and die. Whatever the reason, breeders and entomologists will be unable to pluck resistant varieties from the gene pool. Grain surpluses could quickly turn to grain shortages.

Gary Paul Nabhan of the Desert Botanical Garden in Phoenix, Arizona, an expert on the crops of Native Americans, put it this way: "How we care for the lima bean sprout—or the misshapen maize kernel and the bitter potato—may indeed forecast our own survival."[25]

SEED SAVERS

In 1975, just before he died, a man named Baptist John Ott passed along to his granddaughter seeds of a morning glory, a pink German tomato, and a bean that his family had brought to America from Bavaria four generations earlier. His granddaughter, Diane Whealy of Decorah, Iowa, and her husband, Kent, became fascinated by the so-called heirloom plant varieties, like Baptist Ott's, that families had maintained for generations.

They quickly learned, however, that seeds like Ott's were vanishing. American farmers, like their counterparts in many other places, were rapidly dropping traditional crops in favor of modern, high-yielding varieties. And the garden-seed companies that supplied backyard fruit and vegetable growers were dropping the older varieties from their catalogs. That often meant that the varieties became extinct.

According to the Whealys, seed companies drop about 5 percent of their offerings each year. In 1984, the Whealys determined that there were 239 seed companies in the United States and Canada selling a total of 5,785 vegetable varieties. Just under half of the varieties they sold were available from only one of the seed companies. A total of 3,434 varieties, or 59 percent, were available from only one or two companies.

By 1987, 54 of the 239 seed companies had gone out of business or had been taken over by larger firms. Bankruptcy is likely to mean that the company's unique varieties will disappear. Consolidation often means the same thing as larger, profit-hungry conglomerates drop varieties for which there may be only a small demand.

The Whealys soon became leaders in a movement to preserve heirloom fruit and vegetable crops. "At the farm we're keeping about twenty-two hundred different heirloom beans, seventeen hundred different tomatoes, five hundred different peppers, four hundred and fifty different lettuces," Whealy said. "Our total collection at the farm is probably about seven thousand varieties," stored in airtight jars in their barn. "We grow out about twelve hundred of those each summer, entirely for seed."

The Whealys also maintain a historic apple orchard with about 450 varieties of apples, most of which date from the nineteenth century.

The names of many of the varieties bespeak their folk origins: Pruden's Purple (Potato-Top) Tomato, Moon and Stars Watermelon, and the Stove Wood Bean. The Nancy Watermelon is named after its discoverer, Nancy Tate, who found it in a Georgia cotton field in the 1880s. The Nancy has white seeds and red flesh, and it is unusually sweet and resistant to disease. A cousin of hers took seeds of the Nancy to Arkansas, helping to establish a commercial melon industry there.

The Nancy's downfall was a thin rind. It could not be shipped long distances, and it soon was replaced. Descendants of Nancy Tate's cousin kept the Nancy alive until the 1950s, when they lost it. In 1986, Whealy was able to recover the variety with seeds obtained from Nancy's seventy-eight-year-old son.

Not content to preserve most of their seeds in jars, the Whealys established a network of farmers and backyard gardeners who were willing to grow the varieties, harvest the seeds, and share them with other interested horticulturalists and gardeners.

The Seed Savers Exchange now has six thousand members helping to preserve twelve thousand varieties of fruits and vegetables. Of the six thousand members, nearly one thousand are actively participating in the preservation effort, offering seeds to other amateurs, Whealy said.

"We don't sell seed," he explained. "People write directly to the person maintaining a variety that they'd like to try." Gardeners who do not see themselves as amateur curators are also welcome to join the exchange and raise the fruits and vegetables simply for the pleasure of consuming them, Whealy said. "We've been encouraging it," he said. "A lot of the stories that were published about us made it sound that if you weren't willing to grow and maintain the material, don't even bother with us. We're encouraging people to just experiment with these varieties."

The Seed Savers Exchange was praised in a 1986 report from the Congressional Office of Technology Assessment (OTA), which credited the organization for making a significant effort to help conserve biological diversity. The organization "also has become a focal point for information on rare varieties, a clearinghouse of information on what heirloom vegetable varieties are being maintained at the grassroots level, and where to acquire them," said the report, a survey of grassroots efforts to preserve biodiversity. The exchange was ideally placed to serve as an intermediary between the government and the many individuals engaged in germplasm conservation, the report said.[26]

The Whealys are also beginning a program to conserve animal germplasm. They are raising a cattle breed called Ancient White Park, which was hunted in medieval England but has been reduced to an estimated population of about one hundred. And they are raising Iowa Blue chickens, a vanishing breed once common in northern Iowa.

In 1990, Kent Whealy was awarded one of the so-called genius grants, from the John D. and Catherine T. MacArthur Foundation. Whealy was one of two seed savers honored by the foundation that year. The other was Gary Paul Nabhan of the Desert Botanical Garden in Phoenix.

Nabhan is an authority on Native American agriculture, and he has done much to help conserve Native American crops. While U.S. Department of Agriculture (USDA) plant explorers have combed exotic regions for valuable genetic resources, American genetic resources have been largely ignored, Nabhan argues. In the early 1980s, he and three colleagues established a program similar to Whealy's called Native Seeds/SEARCH. The aim was to preserve native crop varieties from the American Southwest.

The organization stores two thousand varieties of crops, most collected from Native American farmers. Swaps arranged by Native Seeds/SEARCH have restored varieties to Indian reservations that had lost them. Its fifteen-hundred-member gardeners are helping to preserve the varieties in their backyards. And Native Americans are paid to replenish the seeds and encouraged to continue traditional farming practices rather than seek other kinds of work. "The more garden plots those seeds are grown in, the higher the probability they will survive," says Nabhan.

The North American Fruit Explorers (NAFEX), has three thousand hobbyists who do for fruit what Whealy does for vegetables and Nabhan does for the crops of the Southwest. The group holds meetings and workshops to encourage the spread of heirloom fruit and nut varieties. Their collection includes samples not available anywhere else. The OTA's 1986 report singled out the efforts of one NAFEX member, Elwood Fisher of Virginia, who assembled what may be the largest private collection of heirloom fruit varieties in the country. "On only half an acre of land, he has created a preservation orchard containing 840 kinds of apples, 160 pears, 52 cherries, 27 plums, 15 peaches, 47 apricots, 20 grapes, 21 blueberries and many varieties of other fruits and berries—about 2,000 varieties in all," the report said.

The collection assembled by NAFEX is typical of the kinds of germplasm being preserved by backyard gardeners and amateur horticulturalists. The reason their collections often contain varieties not found in government or commercial agriculture is that their aims are different. Government, university, and commercial breeders generally focus on varieties with qualities that make them suitable for commercial production. The varieties must have an attractive appearance, for example, and they must be sturdy enough to survive wearying transcontinental journeys from farm to supermarket. The bland, tough-skinned commercial tomato is a classic example.

Heirloom varieties, on the other hand, are sought for their historical interest, their taste and texture, and other qualities that make them appropriate for cultivation by weekend hobbyists. NAFEX, the Seed Savers Exchange, and Native Seeds/SEARCH are saving varieties that the agronomists have no interest in. In some cases, their research has led to commercial interest in varieties that might have otherwise been overlooked. "To this end, their activities in genetic conservation have served not only the interest of their membership, but, more broadly, the public interest," the OTA said.

MISSING HEIRLOOMS

The heirloom collections of independent seed savers such as Kent Whealy of the Seed Savers Exchange in Decorah, Iowa, are a living historical record of American agriculture. Many of the seven thousand varieties carefully packed into airtight glass jars in Whealy's Decorah, Iowa, barn were once mainstays of American farmers. Most have long since been replaced by tougher commercial varieties, which often sacrifice taste and appearance for hardiness, uniformity, and suitability for machine harvesting.

The OTA, in its 1985 review of grassroots conservation efforts, found that the Agriculture Department was doing little to protect these traditional crop varieties. Without the efforts of Whealy and a few others like him, this living history of American agriculture would be lost.

In addition, the OTA said, the Agriculture Department was doing nothing to preserve the garden varieties of fruits and vegetables that were disappearing from American seed catalogs. As large companies have bought up small seed producers, many of the varieties produced by the small companies have been quietly dropped from catalogs. Without a concerted effort to preserve those varieties, they, too, are in danger of disappearing.

Whealy's collection survives primarily on the strength of his and his wife's commitment. The collection has no institutional support of any kind. The Whealys will not be able to maintain the collection indefinitely. Yet government assistance or interest remains elusive.

"We have very little interaction with the government," said Whealy. "The government collections are focused mainly on large agricultural crops, whereas our collection is focused on fruits and vegetables for home gardeners. The government has enough on its hands trying to maintain what they have, and they don't have much desire to take on ours."

The official neglect of Whealy's work is all the more surprising when the costs are considered. Whealy is preserving seven thousand varieties of fruits, vegetables, and other crops. Each sample is planted and replenished every five years, ensuring against even minimal seed loss. The costs of this expert maintenance—something the Agriculture Department has not been able to come close to—is not millions of dollars or even hundreds of thousands of dollars. Whealy and a staff of two full-timers and three part-timers maintain the collection at a cost of $30,000 per year.

Unlike the government, which is trying to move much of its collection into long-term storage in liquid nitrogen, Whealy believes short-term storage is the answer. "We really don't believe things ought to be stored away in frozen storage. If this material isn't part of the culture, we're all going to become museum keepers presiding over a lot of dead seed. If it isn't something that people grow as part of their life each year, it does become a museum of seeds."

Whealy would like to see his collection duplicated in the U.S. National Plant Germplasm System. "It would give our collections another level of backup, and it would make our collection available to breeders, which is beyond what we're able to do. For those reasons, I'd like to see our material in those collections. But the problems they have with just growing out the material seem to be beyond them."

Henry L. Shands, the director of the U.S. seed-bank system, agrees that backup collections of heirloom vegetables would be desirable. But he does not think it is practical, considering the limits already imposed on government seed banks. Furthermore, he says, the heirloom varieties do not offer breeders as much as wild crop varieties do.

"There's no request for a lot of these old varieties from the 1920s and 1930s," he said. I don't think they [curators] need to keep that in their current collections. They can easily put that in the back room. If there was a delay in getting it, I don't think that would disturb people too much. If you're looking for genes for insect and disease resistance, we've already gone through the old varieties. Now you're looking for new genes. The wild species are the kind of thing that curators are saying will be some of the more useful material in the future."

Whealy sees the matter a little differently. His collection has never been evaluated for disease or insect resistance, he said, so the varieties could be potentially useful in that regard. And the focus of commercial breeding is different, he said.

"Most of the breeding that is being done is commercial breeding, and a lot of the material we have has a different focus. These are garden varieties of use to people who grow their own food. I have friends who are tomato breeders, and if you go to school to become a tomato breeder in this country, you get a Ph.D. focused on growing round, four-inch tomatoes that are machine harvestable. I don't think any of our varieties have traits that are going to help that at all." Whealy's varieties were collected for appearance and taste, not to be tough as baseballs.

The government has not neglected Whealy's point of view. For a time it considered establishing a crop advisory committee to provide guidance on heirloom varieties. That idea was ultimately dismissed. Recognizing Whealy's importance, however, the government appointed him to the bean-crop advisory committee.

"Unlike the government people, I don't get a travel allowance," Whealy said. He has been on the committee for three years, but because of financial constraints he has been unable to attend a single meeting. "We're constantly fighting for funds," he said.

In some cases, the government is doing worse than following a policy of benign neglect. As the OTA pointed out, income-tax regulations and conservation policies sometimes hobble the work of grassroots seed savers.

"These groups frequently serve the public interest at their own expense without expecting reimbursement," the OTA said. But some are threatened, for example, by moves to limit tax deductions for charitable contributions. "Should charitable deductions be reduced or eliminated, donations—a major source of funding—might be seriously curtailed," the OTA said.

In addition, changes in the tax law regarding donation of land and conservation easements—in which land is pledged to be protected from development—are also hurting seed savers. "The frequently complex issues surrounding land donation and acquisition present major obstacles to groups lacking legal expertise," the OTA said.

The OTA praised the grassroots groups for showing how much could be accomplished with limited resources. And it urged the federal government to support these groups.

"The effectiveness of grassroots activities could be enhanced through greater assistance from state and federal agencies in areas that require technical expertise, such as plant or animal breeding or germplasm storage," the OTA said. "Some of these grassroots networks in effect subsidize government responsibilities. And federal or state governments could consider supporting such projects accordingly."

DOCUMENTING THE LOSSES

For all the concern about the desperate condition of America's seed banks, no thorough estimate of the condition of the collections has ever been made. Major Goodman and other critics can only make educated guesses about the status of the national seed collection. The administrators of the system generally agree with Goodman that the resources of the National Plant Germplasm System are far short of what is required to maintain a healthy collection. But many of them feel that the critics went too far. The term "seed morgues," the administrators argue, overstates the case. Yet they are in no position to challenge Goodman's criticism, for they don't have any better information on the status of the seed banks themselves.

Some indication of the extent of the problem came in 1989. In response to requests from the Associated Press, Steve Eberhart, the director of the NSSL, asked his staff to run a computer analysis of the 232,210 seed samples then held in the NSSL. A newly developed computer system had recently made such an analysis possible. The analysis provided data on the "viability" of the samples—that is, the percentage of seeds in each sample that would germinate. As a seed sample ages, its viability gradually declines until the sample is lost altogether. The lower the viability, or germination rate, the more urgent it is to replant the seeds and generate healthy, new seeds.

The results of the computer search were at least as bad as Goodman had suspected. Of the 82,538 samples tested for viability, 77 percent were found to be above the critical threshold level. (Guidelines specify that seeds should be regenerated when the germination rate falls below 65 percent.)

The staggering finding was, however, that the remaining samples—nearly two-thirds of the entire collection—had not even been tested. Part of the reason was that the laboratory's small staff did not have the resources. But the computer search revealed another, more disturbing problem. Thousands of the samples contained too few seeds for the laboratory to risk removing some for testing. Nearly half of the samples in the world's premier seed bank were perilously close to extinction.[27]

Suddenly, Goodman's harsh criticism of the seed banks appeared gentle. He was disturbed about the possible loss of a few thousand corn samples. The computer search revealed that nearly three-quarters of the samples were either untested or too endangered to test.

When the Associated Press published these findings in a series of articles in March 1989, Eberhart objected to the conclusions. It was wrong, he said, to assume that all of the untested samples were in danger. The suggestion that three-fourths of the laboratory's samples might be in jeopardy was a worst-case assessment, and it probably overstated the danger. Besides, he said, many of the samples had arrived at the seed bank in poor shape. The laboratory should not be blamed for allowing them to deteriorate.

He also noted that other U.S. seed banks contained duplicates of nearly all of the endangered samples. But that observation overlooked a central fact. The NSSL was established to be the collection of last resort. It is the secure backup facility whose samples are used to replenish depleted samples at working collections around the country. The working collections are not supposed to back up samples stored at the NSSL. The curators of those collections make their samples freely available to plant breeders on the assumption that any samples that become depleted can be replaced by the NSSL.

But what is the condition of the working collections? Are they in any better shape than the main collection? Again, the status of the collections is difficult to assess. No careful assessment of the collections has been made. Reports prepared for an Agricultural Research Service committee meeting in 1986 suggest, however, that the Fort Collins seed bank is not the only collection that is suffering.

The situation at some of the most important working collections was just as bad. The curator of the nation's principal corn collection, in Ames, Iowa, reported that 35 percent of the samples there were in critical need of regeneration. At a collection in Pullman, Washington, containing beans, garlic, lettuce, lentils, and wild rye, 25 percent of the samples were in critical need of regeneration. The figure for sweet potatoes, peanuts, peppers, melons, and other vegetables in Griffin, Georgia, was 33 percent. In Geneva, New York, 50 percent of the samples were in critical condition. These collections clearly cannot be relied on to back up endangered samples at the NSSL.

The findings published in the Associated Press articles were later confirmed by Joel I. Cohen, a genetic resources specialist at the U.S. Agency for International Development. Cohen and his colleagues found that only about 25 percent of the samples at the NSSL were sufficiently healthy to risk sharing them with other seed banks. And programs for regenerating the seeds were far behind where they should be.

"Without increased funding for regeneration and preservation of collections, there will be no security, and some germplasm repositories may become germplasm morgues." Cohen's criticism couldn't be easily dismissed. Among his collaborators was Henry Shands, the seed-bank system's director.[28]

In 1994, Eberhart again ran a computer analysis to check the status of the collection at the seed-storage laboratory. Eberhart had made great strides in shaping up the collection. By that time, 85 percent of 235,810 samples had been tested to see whether they were still viable. But of the more than two hundred thousand samples tested, nearly one-third were still below the minimum 65 percent germination rate.

Eberhart also reviewed the status of the 443,840 samples scattered throughout the national germplasm system. Each of those samples is supposed to be held in duplicate, at the NSSL and in one of the working collections. The computer check found that 155,034 samples—more than one-third of the entire national collection—was not held in both places. An unexpected plant-disease epidemic, a flood, or a power failure in any one of the collections could wipe out thousands of irreplaceable—and unduplicated—genetic resources.

The situation was particularly bad with respect to fruits and other foods that must be stored as cuttings rather than seeds. The computer check found 27,025 such samples in the system, of which only 307 apple varieties—slightly more than 1 percent of these samples—had been duplicated in Fort Collins.[29] The problem there, however, is only partly due to lack of resources. These samples are much more difficult to preserve. Eberhart's staff has only recently perfected the technology for storing apple varieties. The researchers are close to perfecting the technology for preserving potatoes, citrus seeds, and mint. Even if they are successful, however, they lack the budget to then move ahead and preserve those collections.

The seed-bank systems have been the subject of repeated investigations for more than twenty years by the National Academy of Sciences and two congressional agencies, the General Accounting Office and the OTA. Reports pleading for more funding fill a sizable bookshelf. If the effort put into critiquing the seed banks over the past two decades had been directed toward restoring the system, many of the problems might have been solved by now.

A particularly stinging criticism came in a National Academy of Sciences report released in 1991. The U.S. seed-bank system "is managed by too many individuals, committees and USDA offices," the academy concluded. It found that 50 percent of the samples at the NSSL were below the minimum desired level of 550 seeds.

The response from the Agriculture Department was tepid. "The [academy study] gave good advice to the USDA, which it basically ignored, rejected or could not understand," said Calvin O. Qualset of the University of California, Davis. Qualset oversees California's own seed bank. "There are many excellent aspects of the program, but it just is not placed high enough in the government," he said.[30]

Virtually every agency that has examined the system—including the Agriculture Department itself—has found plenty to criticize. The reports of these investigative agencies have been received with a solemn nodding of heads and little else. Shands and Eberhart understand the problem, but Agriculture Department officials farther up the ladder have stonewalled the criticisms.

Eberhart and Shands have not been the target of criticism.

The laboratory's toughest critics, including Goodman, are quick to praise Eberhart and Shands for their efforts to improve conditions at the laboratory. The point of the criticism is that Eberhart and Shands do not have the resources to do what critics say must be done.

Shands has been more blunt than some of the critics in assessing the problems. "Before germplasm became so exciting a thing, maybe three years ago, it was used as a dumping ground for people who couldn't cut it in pure research," he said.

One former administrator of the small-grains collection, then in Beltsville, Maryland, was dyslexic. He copied plant-identification numbers incorrectly, mislabeling and hopelessly confusing the samples. When shipments arrived, he stacked them in hallways outside the seed vault, unopened, uncataloged, and most importantly, unrefrigerated. Power outages inside the vault were common, depriving the samples of refrigeration.

It is difficult to understand how this reckless handling of the collection could have been allowed, considering the collection's importance. It includes one of the world's most complete collections of wheat, a crop of major economic importance to the United States. The collection was moved to Aberdeen, Idaho, several years later, and Shands took advantage of that opportunity to install a solid administrator.

Shands has also pressed regional seed banks to accelerate their regeneration efforts. "We had a couple of stations where they dragged out their regeneration for twenty years," he said. "I said, 'I don't think that's going to work.' " The seed bank in Griffin, Georgia, is a case in point. It includes collections of sweet potato, peanut, pepper, okra, melons, sesame, and eggplant. In 1990, it contained 63,927 samples. It was replanting fewer than three thousand each year. Meanwhile, the collection was expanding. It was expected to swell to more than one hundred thousand samples by 2001. The regeneration effort was falling behind.

Shands and Eberhart have responded to criticism by pointing to the National Plant Germplasm System's strengths. The storage facilities at Fort Collins are among the best in the world. The testing of the seed samples has accelerated, with thirty thousand samples now being tested each year for germination rate. Recent studies suggest that the seed samples may not be deteriorating as rapidly as Goodman and others feared. An $8.2 million addition to the seed-storage laboratory in 1992 quadrupled its storage capacity.

Samples from the U.S. seed collection are made available freely to researchers in the United States and around the world (as long as the supply of seeds is adequate). The United States shares more seeds than anyone else. For the years 1986–90, for example, the National Plant Germplasm System distributed an average of 192,200 samples each year.[31]

The budget for germplasm increased modestly during the 1980s, but then it stalled. The NSSL's operating budget has been fixed at about $2 million since 1988. The 1992 expansion of the laboratory has siphoned more money into building maintenance, actually shrinking the amount available for germplasm preservation.

"We've asked for about a fifty percent increase," Eberhart said in an interview in July 1994. "A few times, we've gotten some increases through the Agricultural Research Service, and then OMB [the White House Office of Management and Budget] has taken it out." Former agriculture secretary Mike Espy, in response to public concern, gave the highest priority not to germplasm protection but to improving Americans' nutrition and to protecting them from pesticide exposure, Eberhart said. "And yet genetic resources, which are the basis of both of these, are being ignored. If you can use the genetic resources to develop the cultivars, you get less need for pesticides, and you can improve the food quality of many crops."

The main consequence of the static budget is that "we haven't made any progress on doing the germination tests we needed to do. It's difficult to be sure we have high-quality seeds," Eberhart said. The laboratory has also fallen drastically behind in cataloging the new seed samples that continue to arrive daily. "We have them in cold storage. They are being protected. But we haven't tested the germination. They're not available if there is a request," Eberhart said, unable to conceal his frustration. At the time of the interview, the backlog had grown to 28,240 samples.

Seed banks are one of the key sources of germplasm breeders' need to protect the food supply. Even if they are complete and well maintained, however, they do not capture all of the world's biodiversity. Many important breeding resources are yet to be discovered. Iltis's *Zea diploperennis* is one example of the riches still to be found in wild crop relatives and landraces, many of which are best preserved in the field rather than in seed banks. These unusual and exotic plants gave rise to the Green Revolution, which transformed world agriculture in the 1960s. Despite their proven value, these resources, too, are in danger of disappearing.

Chapter 3

Green Gold

During a visit to Japan in 1873, the U.S. commissioner of agriculture, Horace Capron, discovered odd, dwarf wheat plants in the fields of Japanese farmers. He was fascinated by the dwarf wheats, but he did not anticipate that they might have any commercial value. They were simply an intellectual curiosity.

Interest in the Japanese wheats languished until after World War II. In the postwar years, the dwarf wheats suddenly became vitally important around the world, particularly in developing countries. Genes from dwarf wheat varieties were at the core of a spectacular research effort that came to be known as the Green Revolution, which led to an explosion in farmers' harvests. It also provided a stark example of the enormous value that can sometimes be derived from a single plant gene.

The dwarf wheat that fascinated Capron in the nineteenth century was rediscovered immediately after World War II by S. Cecil Salmon, an agronomist with the U.S. Department of Agriculture. He was in Japan as part of Gen. Douglas MacArthur's army of occupation. Salmon's job was to help repair Japan's ravaged economy by getting its farmers back on their feet. While he was there, Salmon sent home seeds from several varieties of dwarf wheats, including one called Norin 10.

Although Norin 10 was new to American wheat breeders, it had some American ancestry. It originated in 1917 as a cross between a Japanese variety called Daruma and an American wheat variety called Fultz, named after its discoverer, Abraham Fultz. A farmer in Mifflin, Pennsylvania, Fultz had found the American variety in his own wheat crop. Daruma and Fultz were not the products of scientific breeding. They were landraces, the product of generations of careful plant selection by farmers like Fultz and his Japanese counterparts. Fultz had no knowledge of the science of genetics, which barely existed when he was saving seeds from the variety that would bear his name. But like other farmers of his day, he had a keen eye for subtle variations in crops.

The Norin 10 that Salmon sent home from Japan was quickly incorporated into American wheat-breeding programs. Orville A. Vogel of the USDA Agricultural Experiment Station in Pullman, Washington, crossed it with a modern variety he had developed called Brevor, producing a Norin 10–Brevor hybrid. Vogel experimented with the hybrid for several years before sending a few of its seeds to Norman Borlaug, a plant breeder at the International Center for the Improvement of Maize and Wheat (CIMMYT), just outside Mexico City.

Borlaug had arrived at CIMMYT in 1944, where he was put in charge of the wheat-breeding program. He was working on what had become a seemingly insurmountable problem for wheat breeders. Breeders had devised wheat varieties that would produce increased yields when doused with increasing amounts of nitrogen fertilizer. The problem was that higher applications of fertilizer caused the wheat plants to "lodge," or fall over. That limited the amount of nitrogen that could be used, as well as the yields.

Borlaug thought that the gene that kept Norin 10 short might alleviate the problem of lodging. Shorter plants would be less likely to fall over. He incorporated the "dwarfing gene" from Norin 10 into the Mexican wheat varieties he was studying. The result was as he had hoped: shorter wheat that could withstand far greater fertilizer applications without falling over. The new wheats were twenty to forty inches tall, compared to fifty or sixty inches for traditional varieties. They had stronger stems and roots, more flowering heads, and more grains per head.

The impact of the new wheats was felt quickly in Mexico. When Borlaug had arrived at CIMMYT, Mexico was producing an average of eleven bushels of wheat per acre. It was forced to import half of the wheat it needed. Twenty-five years later, in 1969, the adoption of the new short varieties had more than tripled the yield to thirty-nine bushels per acre. Mexico became self-sufficient in wheat.

Similar success was soon reported in India and Pakistan.[1] Borlaug's dwarf wheats spread around the world, transforming world agriculture. The achievement earned him the Nobel Peace Prize in 1970. It was this transformation that became known as the Green Revolution.

By 1982, Borlaug's wheats and their direct descendants had spread across 125 million acres of farmland. "The specific use of the dwarfing gene from Norin 10 has affected the food supply of one quarter of the people of the world—one billion plus," said Garrison Wilkes. "And for over 100 million it has been the margin of survival."[2] A single gene from a seemingly unimportant variety of wheat has saved 100 million lives.

DWARF RICE

The Norin 10 story was repeated with rice. The development of dwarf rice varieties began with the discovery in a farmer's field in Taiwan of a dwarf landrace called Dee-geo-woo-gen. Researchers at the International Rice Research Institute in Los Baños, the Philippines (another one of the international agricultural research centers), crossed Dee-geo-woo-gen with an Indonesian rice variety. The result was a dwarf variety they called IR8.

It was introduced to farmers in 1966. Two years later, it had increased the value of Asia's rice crop by $1 billion, although it was being planted on only 5 percent of Asia's rice lands.[3]

The rice revolution was not quite as simple as that achieved with wheat, however. IR8 quickly proved to be susceptible to a variety of pests and diseases. In 1968 and 1969, farmers growing IR8 suffered losses from bacterial blight. The following year, IR8 was attacked in the Philippines by a virus. In the meantime, word was getting back to the breeders in Los Baños that not everyone was happy with the quality of IR8 rice.

IR8 was replaced by an improved dwarf rice called IR20. IR20 also had problems. It was susceptible to attack by a pest called the brown planthopper and to a disease known as grassy stunt virus, which prevents rice plants from flowering and producing grain. Additional strains were developed as breeders struggled to stay one step ahead of the pests in the seesaw battle with nature. The effort culminated in the development, in 1976, of a variety with widespread disease and pest resistance called IR36. By 1982, IR36 covered 27.5 million acres in Asia. It had become the most widely planted variety of rice in history.

The consequences were dramatic. In Indonesia, for example, the switch to IR36 almost doubled rice yields. Once the world's largest importer of rice, Indonesia now grows all of its own rice despite a sizable increase in its population.[4] As was the case with the more recent American battle against the Russian wheat aphid, seed-bank specimens were critically important in the development of IR36. Its ancestry included a wild rice called *Oryza nivara* collected in Uttar Pradesh, India, in 1963. The Indian rice was incorporated into the IR36 genetic mix because it was resistant to grassy stunt virus. Six thousand wild and cultivated rice samples stored in the seed bank at the International Rice Research Institute were screened before *Oryza nivara* was found. It was the only variety with resistance to grassy stunt virus.[5]

The dominance of IR36 will not last forever. In fact, signs of its decline have already appeared. "Even superstars in the varietal relay race eventually succumb to new diseases, pests, or other environmental challenges," said a team of crop scientists, including Donald L. Plucknett, a scientific adviser to CIMMYT and the rice research center. IR36 is becoming vulnerable to new strains of grassy stunt virus as well as the virus that helped doom IR8.

Part of the problem is that IR36 has been planted so widely.

That enormous exposure to environmental assaults will hasten its demise. "The very success of IR36 will contribute to its downfall," Plucknett and his colleagues said. IR60, IR62, and IR64 are replacing it. "Within a decade, IR36 may well be gone, but it has contributed enormously to the feeding of Asia's population. It will stand as a landmark variety attesting to the skill of scientists using genetic resources in exciting and innovative plant breeding, and its genes will form the foundation of future named releases," the researchers said.[6]

The value of dwarf rice varieties has not been limited to the developing world. Rice is one of the four leading cash crops in Texas. The Texas A&M University Agricultural Research and Extension Center, located in China, Texas, has developed eighteen strains of rice worth $2.5 billion to the nation's economy over the past twenty-five years. In 1983, Charles Bollich, the director of research at the extension center, introduced a rice variety called Lemont that was ten to twelve inches shorter than existing varieties. With the adoption of Lemont rice, the cost of producing one hundred pounds of rice dropped from $12.43 to $7.54. By 1988, the new rice had added $1.5 billion to the Texas economy.[7]

The value of crop germplasm, a kind of green gold, is never clear until it is identified and put to use. The examples of Norin 10 and IR36 clearly show that crop germplasm can often be worth billions.[8]

THE VALUE OF WILD CROPS

Green Revolution breeders used two kinds of genetic resources: genes from landraces and genes from the crops' wild relatives. The Green Revolution owed its successes mostly to genes from landraces. Wild-crop relatives are more difficult to crossbreed with domesticated crops. The closer the relationship between two plants, the easier it is to cross them. The wild rice used in IR36, *Oryza nivara,* was a member of the same genus (*Oryza*) as ordinary rice, *Oryza sativa.* That means they were among the most closely related species. Wild crops are often more distantly related to cultivated crops, making it difficult to cross the two.

A variety of other problems have kept plant breeders from fully exploiting the genetic resources of wild-crop relatives. Many wild crop relatives have not been studied to determine their disease and insect resistance. Their seeds are often difficult to collect either because they are produced slowly, over an extended period of time, or because they quickly fall off the plant. Domesticated crops are bred to retain their seeds until the seeds can be harvested.

Despite the difficulty of using wild genetic resources, many breeders believe it is urgent to expand collections of wild crops. Germplasm from the wild crops may be more difficult to work with, but it offers a greater payoff. "Wild plants have been under constant pressure from pathogens, pests, severe climates and unfavorable soils and have evolved myriad strategies for survival," said Plucknett and his colleagues. "It is this arsenal of defensive traits, the product of millions of years of evolution, that is so valuable to agriculture."[9]

But exactly how valuable? In 1986, two environmental consultants, Christine and Robert Prescott-Allen, tried to put a price tag on the value of wild genetic resources. They published their findings in a book entitled *The First Resource,* an analysis of the contributions of wild genetic resources to the North American economy.[10] They wrote:

> Wildlife is the first resource, the exclusive source of food, fiber, fuel and medicines for the first 99 percent of human history. During the last 1 percent, however, wildlife has become the forgotten resource. Agriculture, the use of fossil fuels and industrial development have transformed the human economy, relegating wildlife to a supporting role. Nowhere does this transformation appear more complete than in the industrialized countries, where the primary values of wild plants and animals are widely regarded as aesthetic, emotional, and recreational.[11]

According to their analysis, approximately one dollar of every twenty-two dollars generated in the United States—or about 4.5

percent of the nation's gross domestic product—is attributable to wild species. The most valuable wild species were timber and fisheries. Wild-crop germplasm came third. The Prescott-Allens predicted that wild-crop germplasm will soon surpass fisheries. "Use of wild germplasm for the development of new domesticates and the improvement of established domesticates is the wildlife use most likely to go on growing," they wrote.[12]

The commercial crops that have been improved in part by wild genetic resources are worth about $6 billion per year to the North American economy, the Prescott-Allens concluded. The wild germplasm itself has contributed an estimated $340 million a year to the value of those crops.

Those calculations record only the value of wild germplasm to existing commercial crops. The numbers do not include the value of wild germplasm in providing entirely new crops. The kiwi fruit is the best-known example of a recently domesticated crop. Others include cashews, wild rice, and highbush blueberries.

Blueberries were domesticated in the 1920s and 1930s. They are now the nation's third most valuable berry after strawberries and cranberries. Annual farm sales and imports are worth $44 million. Wild rice—most of which is domesticated, not wild—earns farmers $2.8 million each year. Cashews, grown in the tropics and imported, are the most economically important of the newly commercialized crops. Each year, $116.1 million worth of cashew products are imported. Kiwi fruit is worth about $3 million per year.[13]

The Prescott-Allens' analysis is limited exclusively to wild genetic resources. It excludes genetic resources available in landraces, such as those used to develop Green Revolution crops. Landraces have been used much more widely by crop breeders, and their value to the economy is presumably far greater.

The study by the Prescott-Allens is an exercise in placing a value on biological diversity. Preserving biodiversity is widely seen by biologists as an urgent goal in the late twentieth century as natural habitats come under increasing pressure from population and development. The implicit argument is that biodiversity ought to be preserved because it is critical to the world's food supply—and to human survival.

Not all biologists subscribe to that kind of analysis, however. One critic is David Ehrenfeld, a biologist at Rutgers University in New Jersey. "If I were one of the many exploiters and destroyers of biodiversity, I would like nothing better than for my opponents, the conservationists, to be bogged down over the issue of valuing," he has said.[14]

Economic analyses leave out the intangible values of wilderness, he argued. "We can figure out, more or less, the value of lost revenue in terms of lost fisherman-days when trout streams are destroyed by acid mine drainage," Ehrenfeld said. "But what sort of value do we assign to the loss to the community when a whole generation of its children can never experience the streams?"

The value of endangered species may be even harder to fathom. "If the California condor disappears forever from the California hills, it will be a tragedy," Ehrenfeld said. "But don't expect the chaparral to die, the redwoods to wither, the San Andreas fault to open up, or even the California tourist industry to suffer. They won't." Yet great effort is being expended to try to ensure that the condor will survive. And many people think that the effort is well placed.

"If conservation is to succeed, the public must come to understand the inherent wrongness of the destruction of biological diversity," he writes. "I cannot help thinking that when we finish assigning values to biological diversity, we will find that we don't have very much biological diversity left."

USING WILD GENES

Despite Ehrenfeld's criticism, the Prescott-Allens believe that analyses like theirs can be helpful in public-policy circles. They are trying to counter the arguments of others who point to the economic costs of preserving biodiversity.

Abandoning farms to allow regrowth of forests, for example, costs money—in lost food and jobs. Saving the spotted owl and its old growth forests in the Pacific Northwest is costing loggers their jobs. Arguments about the economic losses of wildlife preservation can be countered with the evidence assembled by the Prescott-Allens. "The contribution [of wildlife] is already being underestimated, because quantitative assessments are not available," they wrote.[15]

More compelling than the theoretical arguments and the economic statistics, however, are the many examples cited by the Prescott-Allens of the way wild germplasm has been used.

The earliest crops to benefit from wild genetic resources were grapes. Commercial grapes were grafted on to wild rootstocks as early as the nineteenth century, before Gregor Mendel began the experimental breeding of pea plants that would lead to the science of genetics. At least 90 percent of California's north-coast grapes are grown on wild rootstocks or rootstocks incorporating wild germplasm. The wild rootstocks were chosen for their resistance to aphids and nematodes. Twenty percent of the grapes grown in the San Joaquin Valley are grafted onto wild rootstocks resistant to pests.

Wild-grape rootstocks have also earned for North Americans the right to enjoy French wines. More than 70 percent of French grapes are grown on rootstocks resistant to the aphid *Phylloxera vitifoliae.* The rootstocks are all derived from wild North American grapes.

A wild grape found in Texas in 1961 has been used to produce new rootstocks for France's Champagne and Cognac regions. Problems with diseases and pests in those regions are compounded by soil that produces an iron-deficiency disease in grape plants. The Texas grape had the only suitable rootstock resistant to the iron-deficiency disease as well as to a fungal disease and the *Phylloxera* aphid.[16]

Another crop saved by wild genetic resources was sugarcane. In the late 1890s, a disease called sugarcane mosaic virus appeared in Java. It spread to nearly every sugarcane-producing region in the world. In 1914 it arrived in Louisiana, where it slashed production from two hundred thousand tons a year to forty-seven thousand tons by 1926.

As is often the case, the critical germplasm with resistance to the virus was found where the virus had originated: in Java. The source of the resistant breeding stock was a wild sugarcane growing on the slopes of Gunung Tjereme, an extinct volcano. Another population of the same wild sugarcane was found in southern India. The two wild crops have been used by breeders to develop commercial varieties of sugarcane with resistance to five major diseases. Without that disease resistance, "there would most probably not be a viable sugarcane industry any place in the world," concluded the Prescott-Allens.[17]

In total, the Prescott-Allens identified twenty-three nontimber crops in North America in which at least 1 percent of annual production is accounted for by varieties containing genes from their wild relatives. The most common use of the wild genes was to boost crop yield by incorporating disease and pest resistance. Increasingly, however, wild genes are being used to improve the quality of commercial crops. Wild germplasm incorporated into iceberg lettuce varieties, for example, have given them sweeter flavor, softer leaves, and an appealing light yellow interior.

Wheat, corn, rice, and barley, which together make up 90 percent of the world's grain harvest, have all been improved by the use of wild-crop germplasm. Bread wheats have been given at least partial resistance to leaf rust and stem rust from wild species. As described above, the incorporation of germplasm from *Oryza nivara* was critical to the Green Revolution in rice.

Two varieties of corn grown in Texas in the 1960s and early 1970s incorporated genes from a wild-corn relative called tripsacum. The wild germplasm boosted yields and helped stop "top firing," the loss of upper leaves during hot weather. Barley was improved with wild stock that made it resistant to powdery mildew, an important fungal disease.[18]

GENETIC SURPRISES

It would be convenient if scientists could determine in advance which germplasm is most likely to be useful to farmers. If the value of germplasm could be predicted, botanists could make intelligent decisions about which areas to explore and which samples to save. Unfortunately, predicting the value of germplasm is like predicting the winner of a horse race. Except that with germplasm, there are tens of thousands of entrants. And the odds are changing all the time.

"A wild resource can be declining into oblivion and then be revived by a sudden change in socioeconomic conditions. . . . The best preparation in the context of wildlife use is to have a safety net of diversity—maintaining as many gene pools as possible, particularly within those wild species that are economically significant or are likely to be," the Prescott-Allens wrote.[19]

In 1925, for example, an Agriculture Department researcher began assembling a collection of wild relatives of sugar beets. Sometime later, he abandoned the collection, and it was nearly forgotten. Fortunately, it was not discarded. In 1983, a new sugar-beet disease appeared in California. The collection of wild relatives, which had been revitalized in 1976, was tested for resistance to the disease. Several resistant samples were found.

Until the new disease appeared, the germplasm stored in that collection was of no apparent value. With the appearance of the disease, the collection suddenly became a uniquely valuable resource. Sadly, the fifty years of neglect had taken their toll. When the collection was examined in 1976, researchers found that half of its samples had died.[20]

Another example occurred with corn. In 1922, an unusual mutant called opaque-2 was found in a farmer's field in Enfield, Connecticut. It was saved by scientists who were interested in the genetics of corn and who kept a collection of rare corn mutants. These so-called genetic stocks are different from either landraces or wild relatives. They are mutants, freak genetic occurrences. Investigation of mutants can shed light on the normal biology of corn, and so they are worthy of scientific study. But such study is considered basic research; the mutants generally have no practical value.

Opaque-2 turned out to be different. Forty years after it was discovered, researchers reported that it had unusually high levels of the amino acid lysine, which is essential for human nutrition.[21] Ordinary corn is low in lysine. The opaque-2 mutant also had high levels of nitrogen and of tryptophan, another amino acid essential for humans.

Opaque-2 was not commercially useful. It produced low yields of poor-quality corn. It was also susceptible to various pests and diseases. But breeders went to work and created new varieties that retained the nutritional superiority of opaque-2 but had higher yields and better-quality grain. These varieties are usually referred to as quality-protein maize.

The development of quality-protein maize has had a dramatic impact on the lives of children in developing countries. Ordinary corn does not contain enough protein to support the normal growth of children. Infants and children getting diets of corn without breast milk or dairy products are likely to be afflicted with diseases of malnutrition, such as pellagra or kwashiorkor. The consumption of legumes with corn lowers that risk, but legumes are often withheld from children, scientists at Johns Hopkins University found.

Dr. George G. Graham and his colleagues at Johns Hopkins tried feeding quality-protein maize to malnourished infants in Guatemala. They found that the corn could support growth like that seen in children given expensive cow's-milk formulas. Such growth is impossible with ordinary corn. "To anyone familiar with the nutritional problems of weaned infants and small children in the developing countries of the world," the researchers concluded, "the potential advantages of quality-protein maize are enormous."[22]

GENETIC EROSION

If American agriculture is to maintain its supremacy, it will need a continued infusion of new, tougher crops. To produce those crops, plant breeders will need new sources of plant germplasm, such as that found in *Zea diploperennis* or opaque-2 corn. Even a casual look at the track record of plant breeding makes that plain enough.

Hundreds, perhaps thousands, of vitally important plants like *Zea diploperennis* remain to be discovered. Botanical expeditions to find those plants—and conservation schemes to protect them—would seem to be a reasonable investment to guarantee a food supply that would meet the needs of the growing population of the United States and the rest of the world.

The fact is, however, that the United States does almost nothing to guarantee the preservation of landraces or wild-crop relatives like *Zea diploperennis.* Researchers know that countless other botanical treasures are out there waiting to be discovered. They know, also, that the natural areas where those treasures might be found are rapidly disappearing, succumbing to an exploding world population and to the spread of urbanization and development. Seeds of *Zea diploperennis* are stored around the world. Its future is secure. But other plants, potentially worth billions of dollars to agriculture, may be disappearing daily. The loss of those resources is a tragic, tangible cost of America's failure to protect the world's biological diversity.

This problem has been apparent for a long time. As long ago as 1936, the U.S. Department of Agriculture acknowledged the problem in its annual yearbook. The department described the barley breeding that had been going on for thousands of years in Asia, Europe, and Africa. "The progenies of these fields . . . constitute the world's priceless reservoir of germplasm. It has waited through long centuries. Unfortunately, from the breeder's standpoint, it is now being imperiled."

The Green Revolution, while it has saved millions from starvation, is now threatening the world's food security. Crop varieties sold by American and European seed companies are spreading across millions of acres of traditional farms around the world. The landraces once grown on those farms are being discarded in favor of high-yielding seeds from U.S. or European seed companies.

The high yield of modern crops is a powerful incentive to farmers to drop their traditional varieties and instead buy seeds from multinational seed companies. Once abandoned by their custodians, the landraces that have been grown for centuries are lost forever. Many of these traditional farming areas remain unexplored by botanists. The landraces have not been preserved in seed banks.

The spread of modern agriculture around the world is destroying the very resources upon which its success depends. Vital genetic resources are disappearing with virtually no notice. "The products of technology are displacing the source upon which the technology is based," said Garrison Wilkes. It is like "taking stones from the foundation to repair the roof."[23]

In a single region, hundreds or thousands of landraces can be replaced by a few varieties of commercial seed. The landraces are abandoned, and the priceless germplasm they contain is lost. In 1936, Agriculture Department scientists warned that if the process continued "the world will have lost something irreplaceable."

"Their voices, as those of all prophets, fell on deaf ears," another scientist wrote in 1971, more than a generation later. "Our problems seemed to be with surpluses rather than deficiencies. As long as the food showed up in the supermarket, nobody cared. . . . The fact that the underdeveloped world contained the germplasm upon which our future depended had not yet sunk in."[24] By that time, much of what the Agriculture Department scientists wanted to preserve had already disappeared.

Now, another generation later, landraces are still vanishing, in what amounts to a colossal hemorrhage of genetic resources. In 1959, farmers in Sri Lanka grew two thousand traditional rice varieties. Today, five modern varieties of rice account for nearly all of the rice grown in Sri Lanka. In India, an estimated thirty thousand traditional varieties of rice have been largely replaced by ten modern varieties that supply 75 percent of India's rice crop.[25]

In Greece, 95 percent of the native wheat varieties have disappeared since the 1950s. José Esquinas-Alcazar of the United Nations Food and Agriculture Organization noted the loss of resources in his native Spain. In 1970, he collected three hundred melon landraces. Four years later, when he tried to replace samples of ten of those, he found that three had already become extinct. The last seeds of a fourth variety were stored in the house of a farmer who had just sold his farm to move to the city with his children.[26]

Wilkes has called this phenomenon "genetic erosion." Landraces are being washed away by a steady rain of modern crops. "The genetic heritage of a millennium in a particular place can disappear in a simple bowl of porridge if the seeds are cooked and eaten instead of saved as seed stock," Wilkes said. Vice President Al Gore, in his best-selling book *Earth in the Balance,* called genetic erosion the single most serious threat to the global food system. "We are, in effect, bulldozing the Gardens of Eden," he wrote.

There are two ways to stem genetic erosion. One is to survey the world's remaining landraces and collect samples to be stored in seed banks. That is called *ex situ* preservation. The other is by encouraging their cultivation in the areas in which they are found, or *in situ* preservation. Preserving landraces "in situ" allows them to continue to flourish and evolve. Putting them in seed banks freezes them in time. They are no longer used and developed, and the knowledge that produced them is lost.

"Traditional or peasant agricultural systems are frequently polycultures that include minor crops and other potentially useful plants," the National Academy of Sciences said in a 1993 report. The weeds that grow among the crops may be an important part of the system, but that relationship—and the knowledge it might provide—cannot be maintained in a seed bank.

In situ conservation is "an important component of the conservation and management of genetic resources," the academy concluded. Preserved areas "can be sources of genetic traits not already captured in *ex situ* collections" and "can also provide living laboratories for studying the genetic diversity of the wild species that are the progenitors of modern crops," the report said.[27]

It would be nice if areas could be set aside, like national parks, to protect landraces from destruction. The problem is that landraces, unlike wild crops, will not survive in untended fields. Cultivated plants depend on human care. They cannot survive on their own. They must be planted, tended, and harvested.

In most cases there is little incentive for farmers to do that when they have access to modern crop seeds. Why should small farmers struggling to feed their families be asked to make sacrifices to preserve a genetic heritage that is important to all of us, rich and poor alike?

Some advocates of in situ preservation have argued that farmers could be offered subsidies to maintain old varieties. But running such a program could be difficult. And it raises ethical questions, as some biologists have noted. "No one has the moral right to coerce farmers to grow low-yielding landraces while others are adopting high-yielding cultivars," one study said. Another problem is the shortage of land for farming in developing countries. In the face of food shortages and hunger, it may be difficult or impossible to spare land for the in situ preservation of landraces. "The subsidy for continuing the planting of unused landraces would have to be sizable to compete with subsistence and cash crops," the study said.[28]

Another suggestion is to pay for in situ preservation of landraces with a levy on sales by commercial seed companies, which ultimately profit from the conservation of breeding stock. But such a subsidy would have to be designed carefully to avoid the appearance that seed companies were paying people to remain peasants.[29] Many observers have concluded that *ex situ* preservation of landraces in seed banks is the only reasonable answer.

SAVING WILD CROPS

The preservation of wild relatives of important crops is in some respects a simpler problem. The critical point that distinguishes wild-crop relatives from landraces is that wild plants do not require human attention. They thrive on their own.

Biological reserves, while unsuitable for the preservation of landraces, are ideal for the preservation of wild-crop relatives. All that is lacking is the political will. With so few people and political leaders aware of the problem, there is no popular demand for the creation of such reserves. Environmental groups, which have the clout to get something done, have never addressed the issue.

The preservation of biodiversity has become an important cause for some American environmental groups. But none of them has drawn the connection between conservation and food.

The preservation of *Zea diploperennis* provides a rare example of how in situ preservation can work. It also illustrates the effort that such preservation requires. The place where *Zea diploperennis* was found is now part of a mountainous area about one hundred miles southwest of Guadalajara known as the Sierra de Manantlán Biosphere Reserve. The reserve was created to protect *Zea diploperennis* and other species from extinction.

The reserve was designed for research, not for recreation. It is accessible by helicopter, but most visitors arrive by way of a grueling daylong drive up a deeply scarred and eroded dirt road from the town of Autlán. The reserve is maintained by a corps of young, dedicated workers, many of them graduates of the University of Guadalajara.

The university spends $300,000 per year to maintain the reserve and operate a research laboratory there. "They have a staff of eighty people on that money—twenty administrators and sixty researchers of all kinds," Hugh Iltis said. The reserve "is the glorystone on the diadem of the university. Everybody thinks it's marvelous."

In addition to the laboratory, there is a bunkhouse that can accommodate two or three dozen visitors, an administrative building, and a large common kitchen where meals are prepared for workers and guests and where the staff congregates in the evenings to read, study, or join in informal songfests. A survey of the reserve has so far tallied 2,500 plant species, more than in any other biosphere reserve in Mexico. The entire state of Wisconsin, Iltis likes to point out, has only 1,750 plant species, although it is nearly ten times the size of the reserve.

Iltis visited the reserve in the company of scientists and a few journalists in December 1988 in connection with a scientific symposium to commemorate the tenth anniversary of the discovery of *Zea diploperennis*. After three days of talks and research reports in Guadalajara, a busload of botanists, including Iltis and Garrison

Wilkes, made the long, difficult two-day drive from Guadalajara to Autlán and from there into the mountains to the reserve.

Two things slowed the trip: the tortuous condition of the road from Autlán to the reserve and Iltis's persistent cries to stop the bus whenever he spied a botanical curiosity beside the road. Iltis is a walking botanical museum, with an encyclopedic knowledge of Mexico's flora and an endless storehouse of information on the historical and scientific importance of cultivated plants, wildflowers, and even weeds.

One plant would trigger a tale about the Spanish conquistador who brought it to the New World. Another would inspire a story about Aztec agriculture or the origin of corn, one of Iltis's favorite subjects. (Wilkes was one of the few on the trip with enough knowledge to vouch for the accuracy of Iltis's fantastic tales. He hinted that, on occasion, Iltis would refuse to let the facts get in the way of a good story.)

Iltis was one of the driving forces behind the creation of the Sierra de Manantlán Biosphere Reserve, and he remains a strong supporter of the need to preserve wild crop varieties in situ. "This 'freezing of the genetic landscape' . . . is the only way to truly preserve the broad genetic base needed," he said. "While agricultural powers pay lip service to this diversity, little in fact is being done to preserve it, while much is being deliberately done to destroy it."

A problem with in situ reserves is that their costs fall on the developing countries of the tropics. That is where the most important plants are still to be found and where in situ reserves must therefore be established. The population of *Zea diploperennis* being protected in the Sierra de Manantlán reserve is of value to breeders, seed companies, and farmers all over the world. Yet Mexico is picking up the tab.

"For accepting that challenge, they deserve to be congratulated," said Wilkes. "The world community is dependent on Mexico." The government's willingness to support the reserve is particularly noteworthy in light of Mexico's huge indebtedness to the developed nations that stand to reap the profits from the preservation of *Zea diploperennis.*

The world is dotted with national parks and other protected areas, but few of them contribute to the preservation of threatened wild-crop relatives. About 3.2 percent of the world's land surface—1.2 billion acres—has been set aside in forty-five hundred parks and reserves around the world. Not all of that area is completely protected. Strict protection extends to about 756 million acres, or 2 percent of the earth's land. (Nearly a quarter of that is in a single national park in a frozen expanse in Greenland.)[30]

Some valuable crop genetic resources might be found in those protected areas. But if so, it is only by chance. In most cases, governments do not even know what crop genetic resources they may have already inadvertently protected.

The protection of crop germplasm is not a priority when government officials are planning reserves. When the boundaries of national parks are being drawn, crop scientists are rarely consulted. Most of the reserves needed to protect threatened crop relatives would be in densely populated developing countries where existing natural areas are often being overrun by squatters and others unable to feed themselves, the crop scientist Otto Frankel has pointed out. Demands for additional reserves specifically designed to protect wild-crop relatives could topple conservation efforts that are already faltering. "It would be a tragedy if genetic reserves of crop relatives were to compete with the all too few nature reserves which protect wildlife as a whole," he said.

In 1980, Robert and Christine Prescott-Allen sent a questionnaire to the government agencies administering protected areas in fifty countries. They were trying to find out how well existing reserves protected wild-crop relatives.

The results were discouraging. "It was clear that by and large protected areas are ill equipped to service the potential users of the genetic resources they maintain," the Prescott-Allens concluded. "Fewer than half of the countries which replied had complied lists of species occurring in their protected areas—and then only for a small minority of the reserves. For the time being, therefore, it is impossible to tell how many wild relatives of crops occur in protected areas, and consequently it is impossible to decide whether new areas are needed and, if so, where."

In the absence of a determination of the species that dwell in a given reserve, botanists cannot tell what is protected and what isn't. It is impossible to be sure that wild-crop relatives are indeed being protected. "It may be that [a] reserve is ostensibly protecting populations of wild cereals or wild grapes, but that in practice those populations are being overgrazed by herbivores that the reserve is also attempting to maintain," the Prescott-Allens said.

A further problem is that many protected areas are set up with the understanding that nothing can be removed from them. Even if researchers could locate plants of interest, local conservation regulations and prickly international relations often prevent researchers from collecting germplasm for study or preservation. Less than 15 percent of the countries that answered the Prescott-Allens' survey permit the collection of germplasm in their reserves. "At present," the Prescott-Allens concluded, "in situ gene banks are more a hope than a reality."[31]

U.S. BIODIVERSITY PRESERVATION

As with the seed banks, it would be nice to hold up the United States as an example of how things should be done. American efforts to preserve its own species in situ have been dismal. Wild species in the United States are threatened by the same loss of habitat that is occurring in the tropics. The United States has fewer wild-crop relatives to protect than tropical countries do. But even with a much simpler job to do, it has failed to protect its native-crop species. One example is the failure to protect the wild rice that has contributed $1.5 billion to the economy of Texas.

Fifty years ago, Texas wild rice, *Zizania texana,* was a common sight along the San Marcos River and in irrigation ditches near San Antonio. As the population of the area grew, however, the rice suffered. A decade ago, it survived only in a small patch of about a thousand square yards in the river, according to Gary Paul Nabhan, a biologist at the Desert Botanical Garden in Phoenix, Arizona. The damming of the San Marcos River had raised the level of the river to the point where nearly all of the wild rice was submerged. Once under water, the rice was unable to reproduce.

A tiny bit of rice survived inundation. But it is now threatened with the loss of water. Increasing urban growth around San Marcos is exhausting the spring that feeds the rice. If the spring dries up, the rice will have no chance of survival. The problem has not gone unnoticed. Faced with the imminent extinction of Texas wild rice, biologists have begun a search for an area suitable for in situ preservation of Texas wild rice. But where? Nearly all efforts to transplant Texas wild rice somewhere else have failed.

Seeds preserved in water tanks in greenhouses died. A few plants transplanted to the Salado Creek in Texas were destroyed by bulldozers and boaters. Transplants in the Comal River were destroyed by floods. Transplants in other parts of the San Marcos River were accidentally destroyed by river-sports enthusiasts in canoes and tubes. There is noplace left to transplant it.

"Even if *Z. texana* habitats were to be more fully protected, it would be difficult to restore them to their original conditions," Nabhan said. The rice is in a population bottleneck. Only a few individual plants remain. Much of the original genetic diversity within the species has already been lost. Even if some way is found to preserve the wild population, that genetic diversity will not come back. It would be like trying to reconstitute the world's human population, in all its current variety, from the offspring of a few hundred people in San Antonio. It simply can't be done.[32]

Texas wild rice is not an isolated example. The sunflower is another important crop native to the United States. About a dozen species are endangered. That is one-fourth of all the species that remain in the United States. Several have already disappeared. "The remaining rare species receive little protection in habitat; at best, a few seeds are stuffed into an envelope and considered 'saved,' " said Nabhan.[33] Wild sunflowers have been used to breed numerous disease-resistance genes and other desirable traits into commercial sunflowers and their cousin, the Jerusalem artichoke.

The Agriculture Department has collected seeds of the remaining sunflower varieties for preservation *ex situ* in seed banks. Despite the advantages that *in situ* preservation can provide, however, the U.S. Department of Agriculture (USDA) has done little to encourage it. "For some reason, it has shied away from advocating *in situ* conservation of plant genetic resources, even though the USDA administers many lands that harbor species needed by breeders," Nabhan has written. "USDA publications even claim that *in situ* conservation is too unstable and lacking in long-term continuity to be a worthwhile investment."[34]

Many other U.S. plants face a crisis similar to that of wild rice and sunflowers. The Agriculture Department's Plant Exploration Office has determined that at least 37 of the 250 species now listed as endangered or threatened under the Endangered Species Act carry genetic traits potentially useful in crop breeding or horticulture.

Nabhan's criticism of the government extends to the National Park Service, steward of many of the nation's most important natural areas. The park service has done little to assess or to protect the genetic resources in national parks. "Some national park administrators are even reluctant to allow the collection of seeds of known plant genetic resources for backup storage and evaluation in seed banks and botanical gardens, even when there is no danger that this collection will deplete the plant populations," he said.[35]

"In terms of understanding the plants, animals, fungi, and microorganisms of the United States, we have a very long way to go," said Peter Raven, the director of the Missouri Botanical Garden in St. Louis. "I estimate that there may be some 250,000 species of organisms living within our boundaries, yet no more than 150,000 of these have been identified to date," he told a congressional subcommittee in 1991. "At a time when Costa Rica, Taiwan, and other areas of the world have effective national biological surveys, the United States remains without one."

An indication of the number of species still to be discovered within U.S. borders has come from studies of insects, spiders, and mites in the soil of the old-growth forests of the Pacific Northwest. At a single forest research station in Oregon, scientists found some eight thousand distinct species, most of them in the soil. The finding was surprising, because only tropical forests were thought to have that degree of biodiversity.

About thirty-four hundred of the eight thousand species were arthropods (insects, crustaceans, or spiders). Many of them were new to science. Despite the enormity of the discovery, John Lattin of Oregon State University estimated that researchers had uncovered only about half of the species living at the study site. Investigators do not know why so many thousands of species coexist at the Oregon research site. But the invertebrates would not be there if they were not performing some essential role in the forest ecosystem.[36]

Arthropods, of course, are not useful in plant breeding. But the lesson is clear. Extraordinary biological wealth still remains to be found, even in the United States, in the most familiar and seemingly well explored areas. The arthropods discovered in the Northwest might prove useful in combating insects that attack crops. If such natural riches are not tallied and investigated, no one will know. And they may disappear before anyone finds out.

In the spring of 1989, the Keystone Center in Keystone, Colorado, held a series of meetings to determine how well the United States had cataloged the biological wealth on federal lands. The Keystone Center is a nonprofit organization that searches for consensus in contentious public-policy debates. Participants in the meetings included government officials, environmentalists, industry representatives, and academics. In a report issued in April 1991, the participants concluded that the government's efforts to tally its resources were disastrous.

The Bureau of Land Management, for example, is required to periodically inventory its lands. It is one of the most important government landholders. The agency manages 270 million acres of public land—about one-eighth of all the land in the country. The bureau's land is supposed to be managed for wildlife conservation as well as recreation and development. Yet the Keystone participants found that the bureau had little idea what kind of wildlife resources it controlled.

Rangeland vegetation maps were completed for only 25 million of the 270 million acres. The bureau had devised an Integrated Habitat Inventory and Classification System to survey wildlife resources on its lands, but less than 5 percent of the bureau's lands had been inventoried. A "Threatened and Endangered Species Data System" had been developed to store information on vanishing species. The system contained data on less than 1 percent of the threatened and endangered species on bureau lands.[37]

The U.S. Fish and Wildlife Service, which is likewise directed to conserve wildlife, had also fallen far behind in tallying the resources on the 90 million acres it controls. Waterfowl and endangered species were the focus of most inventories on Fish and Wildlife Service lands. Plants, some of which may be of interest to crop breeders, receive limited attention.[38]

The 191 million acres administered by the U.S. Forest Service had been more carefully inventoried than most other federal lands. But most of the inventories were designed to provide information for logging. The national forests and national grasslands under the jurisdiction of the U.S. Forest Service are of considerable biological importance. They contain 70 percent of the major vegetation types found in the United States, along with substantial representation of other wildlife. About 171 threatened or endangered species are known to exist in the national forests. The Forest Service is actively working on only twelve of them. The spotted owl of the Pacific Northwest is receiving more money and more attention than all of the other threatened and endangered species combined.[39]

The National Park Service, custodian of 80 million acres, is in better shape, particularly with regard to plant species. For some national parks, 75–95 percent of the plants and other species have been tallied. But the inventories are not adequate either for dealing with emerging problems or predicting future problems, according to the Keystone report.[40]

In 1993, Secretary of the Interior Bruce Babbitt called for a national biological survey that would remedy some of these deficiencies. He asked a committee of the National Academy of Sciences to make proposals for such a survey. In a detailed report released in October 1993, the committee listed the reasons for conducting such a survey and spelled out how it should be done. The report strikingly overlooked agricultural resources.

It demonstrated once again that even leading ecologists fail to recognize the importance of conservation to the U.S. food supply. The committee was led by Peter Raven, one of the country's leading ecologists and conservationists. It included eighteen distinguished scholars but none of the leading authorities in germplasm conservation. The report talked about preserving the nation's biological heritage, preparing for global warming, and safeguarding the water we drink and the air we breathe. It said virtually nothing about protecting America's food.[41]

MISSED OPPORTUNITIES

During the 1980s, the issue of preserving the world's biodiversity began to attract some attention in the United States and Europe. But biologists did not do a very good job of selling the rationale. Why should threatened and endangered species be preserved? Even advocates of the idea seemed to have difficulty explaining it. The motivation often appeared to be primarily recreational: to preserve wilderness for hikers and campers.

Sometimes an ethical reason was offered: The human species should not wantonly drive other species to extinction. "The sad fact that few conservationists care to face is that many species, perhaps most, do not seem to have any conventional value at all," said David Ehrenfeld. "True, we cannot be sure which particular species fall into this category, but it is hard to deny that there must be a great many of them."[42]

Sometimes environmentalists simply fell back on the Mount Everest justification: Wilderness should be preserved because it is there. For those less spiritually inclined, the notion of the natural pharmacy was sometimes advanced. Plants have been the source of about one-fourth of prescription drugs. We should not destroy any more plant species, the argument goes, until we have examined what pharmaceutical wonders they might contain.

All of those justifications for preserving biological diversity have their adherents, and all are reasonable. Throughout the ongoing discussion of the issue, however, one justification for conserving biodiversity was conspicuously absent. That is the role of biodiversity in preserving the earth's habitability—by assuring that people have something to eat.

Campaigns to preserve biodiversity often focused on animals rather than plants. Conservation groups built fund-raising and public relations campaigns around so-called charismatic species—endangered animals that know how to perform in front of television cameras. Whales, pandas, mountain gorillas, and Jane Goodall's chimpanzees are examples. While the campaigns focused on individual animals, their real aim was often to protect the ecosystems in which the animals live.

The fight over the spotted owl in the American Northwest is the classic American example. Conservationists see the protection of the owl as symbolic of the protection of the entire ecosystem of the old-growth forests of the Northwest. But why should the forests be preserved? Again, the answers failed to take note of the role of biodiversity in allowing the earth to support human life.

The argument shifted slightly in the late 1980s, with the discovery of the drug Taxol. Isolated from the Pacific yew tree, Taxol is proving to be a potentially important drug for the treatment of ovarian cancer. It was approved by the U.S. Food and Drug Administration in December 1992—the first one to require an environmental impact statement. If the old-growth forests had been destroyed by logging, Taxol would not have been discovered. Do the ancient forests of the Northwest likewise contain plants that could be useful in crop breeding? Maybe not. But before Guzmán and Iltis discovered *Zea diploperennis* in Mexico, no one knew whether the Sierra de Manantlán mountains contained any useful plants, either.

The habitat of *Zea diploperennis* lacks charismatic species. There are no whales or pandas in the Sierra de Manantlán mountains, nothing that could attract tourists or anchor a fund-raising campaign.

To the casual visitor, the Sierra de Manantláns are a far less attractive lure than nearby Acapulco. The mountains are attractive, but they are not scenically striking enough to build a fund-raising campaign around them. Yet the mountain habitat is precisely the kind of place that is home to many of the world's most valuable traditional crop varieties and wild crop relatives. If a photogenic endangered animal or the scenic splendor of a Grand Canyon or a Yosemite is what it takes to get an area protected, the wild-crop relatives that have so far escaped destruction are certainly going to disappear.

The only alternative is to collect the seeds of important crops and preserve them *ex situ:* in seed banks. Threatened areas could be identified. Platoons of botanists could move in with pruning shears to harvest samples of all the important wild-crop relatives. The problem with collecting seeds of wild plants and storing them in seed banks is that nobody knows for certain what to collect. Which plants are likely to contain genes useful in crop breeding? It takes a detailed, time-consuming study to answer that or even to make intelligent guesses.

Furthermore, even if botanists did know what to collect, preserving seeds in a seed bank is not a substitute for preserving plants in the wild. Living, growing plants in wild areas or in farmers' fields are constantly confronted by new pests and diseases, in the very same way that commercial crops are. Commercial crops are protected by disease and pest resistance bred into them, supplemented by the application of pesticides.

Wild-crop relatives have no such protection. They must evolve their own defenses if they are to survive. If they are stored in refrigerated vaults, they are no longer subject to the forces of natural selection. They are not exposed to new pests and thus cannot develop defenses against them. Without those defenses, seeds of wild-crop relatives could be vulnerable to new pests or diseases as soon as the seeds are removed from seed banks.

"I can imagine a scenario in which the rare, valuable plants now sheltered in seed banks, once released, succumb to newly arisen problems that did not exist when they were put into storage," Nabhan wrote.[43]

Consider, he said, the problem of the increasing saltiness of agricultural land in the West. The salinization of the soil is a consequence of excessive reliance on irrigation. Soil flushed repeatedly with water containing small amounts of salt tends to hold on to the salt. Salty soils are leading to reduced crop yields in some areas, and the problem is growing. Breeders are experimenting with salt-tolerant crop varieties, which may someday be the only kind that will grow in many parts of the world.

"Let us imagine that at that time, a crop breeder searching for salt-tolerant plants discovers that a twentieth-century scientist once banked seeds of a wild bean relative that grew on the saline playas of the Mexican coast," Nabhan said. "That wild bean population is now extinct in the wild, and no other sources of salinity tolerance are known for cultivated beans."

Assuming that the seeds have been properly cared for in seed banks, a future breeder can request a sample. He can then grow the beans in his greenhouse, where he can use as yet unimagined tricks of biotechnology to transfer the salt tolerance to commercial crop varieties. But, said Nabhan, what if the plants become infested with an insect-borne virus that did not exist when the seeds were locked up in the seed bank?

Air pollution, acid rain, or unusual soil conditions that did not exist in the twentieth century might likewise doom the new crops. In Nabhan's scenario, the prized salt-tolerant genes could be "rendered meaningless by the altered conditions they find on the ground a century later."

Nabhan, the recipient of a MacArthur Foundation "genius grant," believes strongly in the value of in situ preservation of plant genetic resources. He has also urged that more attention be paid to preserving the culture and practices of Native American farmers along with their valuable traditional crops. "As a kind of survival insurance, seed banks may be fine," said Nabhan, "but there will be tremendous losses if we assume that they are all we need in the way of long-term conservation measures."[44]

Edward O. Wilson of Harvard University noted another difficulty with seed banks. "If reliance were placed entirely on seed banks, and the species then disappeared in the wild, the bank survivors would be stripped of their insect pollinators, root fungi, and other symbiotic partners, which cannot be put in cold storage."[45] Each crop or wild relative is part of a tiny ecosystem that a seed bank cannot possibly preserve.

Iltis has gone even further, arguing that seed banks are not only incapable of doing the job but that they aggravate the problem. According to Iltis, seed banks are "an anti-solution, instilling a false sense of security. Not only is such stored diversity limited, but it is open to accidents of technology, such as power failure, war, lack of human foresight, as well as genetic changes inherent in the storage process itself."

One intermediate solution that combines some elements of in situ preservation and some aspects of seed banks is the preservation of plants in botanical gardens. Like in situ reserves, botanical gardens allow plants to continue to adapt to changing conditions. Like seed banks, botanical gardens give researchers easy access to carefully stored and cataloged specimens. "Botanic gardens are extremely important tools for maintaining species and genetic diversity," the Congressional Office of Technology Assessment said in 1989.

Collections in botanical gardens are limited in size, however. The world's fifteen hundred botanical gardens contain about thirty-five thousand plant species, representing 15 percent of the world's plants. England's Kew Gardens contains an estimated twenty-five thousand plant species, about twenty-seven hundred of which are rare, threatened, or endangered.

Many of the gardens contain a high percentage of the flora in their region. One in California, for example, contains one-third of California's native species. Nineteen U.S. botanical gardens have allied to form the Center for Plant Conservation, which is intended to preserve the three thousand species native to the United States that are now in danger of extinction. More than three hundred of them are cultivated in the allied botanical gardens. But botanical gardens, like seed banks, cannot be a substitute for preservation of plants in the wild.

Although the United States has done part of the work necessary to safeguard its agricultural resources, it has failed to use those resources to protect its crops. Only a handful of the seed samples in the vast collection at the NSSL have actually been used to improve U.S. crops. The result is that U.S. crops are more vulnerable to epidemics than they should be. And that is not just a theoretical concern. Two decades ago, U.S. corn farmers found out how real it could be.

Chapter 4

"As Alike as Identical Twins"

In the middle of the nineteenth century, Ireland was the most densely populated country in Europe. In 1841, census takers put the official population of Ireland at 8,175,124. The actual population was probably slightly higher. Unlike the rest of Europe, Ireland was a land of family farms. The verdant Irish countryside was almost entirely carved up into small, separate fields. Sixty-six percent of the Irish people depended on agriculture to earn a living.[1]

In most places, the Irish soil was too poor to support grazing animals. Instead, Irish farmers had turned to the potato, an import from South America. Potatoes took well to the Irish soil. They produced a relatively large amount of food from a small plot. That made them ideal for feeding the large and growing Irish population.

Because they were so well suited to conditions in Ireland, potatoes were adopted by nearly all of Ireland's farmers. "The existence of the Irish people depended on the potato entirely and exclusively," wrote Cecil Woodham-Smith in *The Great Hunger.* Varieties planted in Ireland included the black potato, the Irish apple, the kidney potato, and the red potato. But the most widely grown variety was the lumper. All of the lumpers grown across Ireland were genetically identical.

The Irish had no illusions about the dangers of total reliance on a single crop. Nearly two dozen crop failures had been recorded by the middle of the nineteenth century. In 1728, census commissioners recorded "such a scarcity that on the 26th of February there was a great rising of the populace of Cork." In 1739 and again in 1740, the crop was entirely destroyed. It failed again in 1770 and 1800. About half of the crop was lost to frost in 1807.

Scarcely a decade went by in the early nineteenth century without several major crop failures. "The unreliability of the potato was an accepted fact in Ireland, ranking with the vagaries of the weather, and in 1845 the possibility of yet another failure caused no particular alarm," wrote Woodham-Smith. Those crop failures, however, did not forecast what was to come.

The year 1845 promised to be a good one. On July 23, Irish authorities reported that the potato crop had never been as large and abundant. Two weeks later, Sir Robert Peel, the British prime minister, received a letter from the Isle of Wight. It advised him that potato blight had been found.

Potato blight was caused by an infestation of the so-called late blight fungus, *Phytophthora infestans*. The blight was known in North America; its detection off England was the first indication that it had crossed the Atlantic. (Earlier Irish crop failures had been caused by weather and other diseases, such as dry rot.)

By August 23, reports were suggesting that the blight had moved beyond England to Ireland and was spreading across Europe. "A fearful malady has broken out among the potato crop," wrote John Lindley, a professor of botany at the University of London. "On all sides we hear of the destruction. In Belgium the fields are said to be completely desolated. . . ." According to Lindley, botanists had no way to prevent or eliminate the blight. On September 13, Lindley reported that the blight had reached Ireland.

Irish farmers hoped that the crop that had seemed so promising only a month before might somehow survive the epidemic. In October, however, the potato harvest began, and that hope vanished. The crop was almost completely destroyed. The lumper variety had turned out to be vulnerable to the blight.

Most of Ireland survived that difficult year. The devastation came the next year. Farmers had no choice but to plant the same potatoes again. They had no other varieties. The blight struck again, this time with overwhelming force.

The suffering was indescribable. The next census showed that Ireland's population had been reduced by 2 million. Because of inaccuracies in the census counts, historians cannot say for sure how many people died. But by one estimate 12.5 percent of the population, or about 1 million people, perished from starvation. Another 19 percent, or about 1.5 million, emigrated, most to the United States. The rest were left in crushing poverty, with the Irish economy in shambles.

Although the blight had spread across Europe, its toll was far higher in Ireland than elsewhere because of Ireland's almost total reliance on potatoes. Political conditions in Ireland further magnified the blight's effect. "The horrors of the Irish famine were compounded by the ignorance and inhumanity of the absentee landlords, the restrictive English Corn Laws, the lack of food distribution channels in the countryside and the extreme severity of the winters of 1845 and 1846," the National Academy of Sciences study on genetic uniformity said.

The potato is a native of South America. It was domesticated in the Andes eight thousand years ago. By the time Europeans arrived in the New World, hundreds of potato landraces had been developed. But only two samples were taken to Europe: one to Spain in 1570 and another to England twenty years later.

The potato did not win immediate acceptance. Its spread was encouraged in France by one Antoine-Auguste Parmentier. As a prisoner of war in Hannover, Germany, in 1757, Parmentier survived only by eating potatoes. Eager to promote potatoes in his homeland, he persuaded King Louis XVI to loan him a field near Paris. He then asked the king to station royal guards around the perimeter of the field by day but to leave it unguarded at night.

As Parmentier had hoped, local farmers crept into the field at night to harvest samples of the exotic crop the king evidently valued so highly. Potatoes soon sprouted throughout the country. When Parmentier presented the king with a bouquet of potato blossoms, Marie Antoinette put one in her hair.[2]

At the time of the Irish potato famine, virtually all of Europe's potatoes were derived from the two samples that had been brought back from the New World in the sixteenth century. Unfortunately, both samples were susceptible to infection with the fungus that caused potato blight. All of Europe's potatoes were at risk.

At the time, botanists did not know what caused the blight. Many blamed the foggy, rainy weather during the peak epidemic years. They observed fungus on the leaves of the potato, but they also saw that the fungus did not appear until after the leaves had begun to wilt from the damp weather. They decided that the fungus was a consequence of the potatoes' destruction rather than its cause.

They were wrong, but they did make a valuable observation. Variations in weather can affect epidemics. The rainy weather had favored the spread of the fungus. In dry years, the damage to the potato crop was much less severe.

Initially, Ireland had been protected by geography from the ravages of *Phytophthora infestans*. When the fungus crossed the Atlantic and overcame the geographic barrier, no genetic barriers to its spread remained. The picture was quite different in the Andes. Farmers there grew hundreds of different potato landraces. Some were damaged by the blight; some were not. So at least part of each farmer's crop was safe. There was no epidemic.

That genetic diversity was not available to Irish potato breeders. "The early potato breeders found themselves like card players with only a tiny part of the deck," wrote Robert and Christine Prescott-Allen.[3] Some new potato varieties from South America became available in 1851, but none were resistant to the blight.

In 1908, a Mexican potato relative that hybridized naturally with potatoes was discovered to carry blight resistance. More resistant varieties were found in 1925, when Nikolai Vavilov dispatched a botanist to Central and South America to collect wild and domesticated potatoes. The wild-potato relatives were widely used to make European potatoes more resilient. Of 586 potato cultivars grown in Europe, 320 carry genes from wild-potato relatives, according to the Prescott-Allens. The potato is once again a staple crop in Ireland.

THE GENEALOGY OF THE CORN BELT

In the American Midwest, corn plays the role that the potato played in nineteenth-century Ireland. Corn arose in southern Mexico and Central America more than seven thousand years ago. (It is the only major U.S. crop to have originated in North America.) Long before the arrival of Columbus, corn was being grown across North America. Its exact origins are unknown. It is known to be a relative of teosinte, the wild cornlike plant found in southern Mexico and Central America, where corn originated. But the exact relationship between teosinte and domesticated corn has been the subject of a scientific debates so fierce that some botanists have referred to them as the "corn wars."

The principal combatants in these academic contretemps were Paul Mangelsdorf of Harvard University, and George Beadle of the University of Chicago. As the writer E. J. Kahn, Jr., tells it in his book *The Staffs of Life,* Mangelsdorf believed that teosinte had always been a cousin of corn—that the two had evolved in parallel. Corn's direct ancestor was some kind of corn, not teosinte, according to Mangelsdorf. The ancestor of corn had either disappeared, or it had not yet been discovered, Mangelsdorf argued.

Beadle thought otherwise. He was convinced that teosinte was the ancestor of corn. The two men waged an angry academic battle for decades, with neither ever yielding much to the other. Garrison Wilkes, a student of Mangelsdorf's, wrote a doctoral dissertation critical of Beadle's views but never came down squarely on either side of the dispute. "I have been criticized on occasion for not committing myself to one of the competing theories for the origin of maize," Wilkes said. But he said that he had not seen what he considered irrefutable evidence for either proposition.

As botanists continued to search for archaeological evidence or other clues that would allow them to draw the correct family tree for teosinte and corn, Hugh Iltis—the discoverer of perennial teosinte—promulgated an entirely different explanation for corn's origin. Iltis, with typical flair, announced what he called the "catastrophic sexual transmutation theory."

Corn plants carry both male and female parts. The ear is the female part; the tassel that crowns the plant is the male part. Teosinte, unlike corn, has multiple branches, each with its own tassel. Iltis proposed that some of the male tassels of teosinte had undergone some chance transmutation to female ears. These small, primitive-looking teosinte "ears" then evolved into modern ears of corn as farmers, for thousands of years, selected plants with the largest ears.

Modern genetic analyses of corn and teosinte have still not put the matter to rest. Part of the problem is that corn and teosinte are so different, said Major Goodman of North Carolina State. "I can't think of another major crop that has such a gap between the cultivated form and the putative wild ancestor," he said. "It's a Grand Canyon that separates the two, and that's what caused so much controversy over the years."[4]

The debate has shifted, however, with most researchers now accepting that corn somehow evolved from teosinte. "In recent years, a large body of evidence has accumulated that supports the hypothesis that maize is a domesticated form of teosinte," John Doebley of the University of Minnesota and his colleagues wrote in the *Proceedings of the National Academy of Sciences* in December 1990. "Few authors still question this view," they said.

J. Stephen C. Smith, a corn geneticist at Pioneer Hi-Bred International in Johnston, Iowa, agrees. "The bulk of evidence would support that teosinte could have given rise to corn. But it's not proven."

Whatever the ultimate origin of corn, archaeologists have determined that varieties similar to modern corn were being cultivated in Mexico seven thousand years ago. Columbus first saw it in November 1492, when two of his crewmen presented him with corn they had been given in the Cuban interior. Native Americans had developed a number of corn varieties that they passed on to the European colonists.

Principal among these was the New England flint variety, a

slim ear with eight rows of broad kernels. The flints were cultivated by the earliest colonists. In Virginia and farther south, farmers grew a race of corn called Gourdseed, or southern dent, with many rows of thin kernels. America's westward expansion was to unite the two in a marriage that would give birth to the Corn Belt.

"As settlers pushed west through the Cumberland Gap, they took the southern dent types from Virginia, with which they were familiar, and planted these on the lands newly cleared of forest," said Wilkes. "Sometimes the germination was poor and the empty spots were planted with the more rapidly maturing New England flints. The two races both flowered at the same time. Hybridization took place, and a unique American race was formed: the highly productive corn-belt dent. The whole process took about 150 years."[5]

Until the twentieth century, improvements in corn were made by selection: Farmers saved seeds from the plants that were hardiest, the most productive, or produced the most desirable ears. As early as the late 1870s, however, researchers began to explore the possibilities opened up by making deliberate crosses of one variety of corn with another. Hybrid corn was born.

Corn Belt farmers were initially resistant to the new varieties produced by these deliberate crosses. But when seed companies began demonstrating superior yields with hybrid corn, farmers embraced them seemingly overnight. By the end of World War II, hybrids had almost completely replaced the so-called open-pollinated varieties that had preceded them. The widespread acceptance of hybrid corn encouraged a dangerous genetic uniformity in the U.S. grain belt that would ultimately lead to a disastrous epidemic.

THE RISE OF HYBRID CORN

One of the first researchers to produce and study hybrid corn was William J. Beal of the Michigan Agricultural College, who began his experiments in the 1870s. According to Deborah Fitzgerald, a Massachusetts Institute of Technology historian who tells the story in her book *The Business of Breeding,* Beal chose corn because it had features that made it ideally suited for crossbreeding.

Unlike many other plants, corn has separate male and female parts. The male, pollen-producing parts are all in the tassel. Borne by the wind, the pollen fertilizes the female ear that emerges alongside the stem of the plant. Corn silk is the conduit that carries pollen inside the husk to the developing kernels. Each individual kernel has its own silk "pipeline" to the wind-borne pollen outside the husk.

Beal began his crossbreeding experiments by planting two varieties of corn in alternating rows. The idea was to have the pollen from the male tassels of one variety fertilize the ears of the other. Beal would decide which variety was to be the "male" parent, providing the pollen. He would then remove the tassels from all the plants of the other variety. Only one of the two varieties, then, released pollen. When it fell on the ears of the other variety, the result would be a cross between the two. (The variety being used as the male would pollinate its own ears as well. They would not be hybrids and thus would not be part of the experiment.)

Beal's hybrid corn plants had yields 10–50 percent above ordinary open-pollinated varieties. By the 1880s, Beal's technique had been adopted by other researchers, although it was not adopted by commercial breeders. The rediscovery of Mendel's laws of genetics after the turn of the century accelerated these efforts.

At first, though, Mendelian genetics seemed an important theoretical concept with little practical application. George Shull, a researcher at the Cold Spring Harbor Laboratory on Long Island, chose corn as an experimental subject for the study of Mendel's laws. Between 1904 and 1911, Shull conducted experiments that led to a rudimentary explanation of the increased yields Beal had seen in hybrid plants. The phenomenon is known as heterosis, or hybrid vigor.

Hybrid corn plants demonstrate this so-called hybrid vigor in several ways. A hybrid offspring may be taller than either of its parent varieties. The leaves, roots, or individual cells may be larger. Or, most importantly, the ears and kernels may be larger. The process is not entirely understood. One theory is that the offspring inherits favorable genes from each of its parents and has more favorable genes than either parent alone.

The hybrid offspring also inherits unfavorable genes from each parent. But the assumption is that favorable genes are dominant, overpowering the effects of recessive unfavorable genes—that is, if a plant inherits a favorable gene for a given characteristic from one parent and an unfavorable gene from the other, it displays the favorable characteristic. Corn that is self-pollinated, or inbred, the theory goes, increasingly acquires pairs of recessive, unfavorable genes for various characteristics with each generation. So its overall vigor declines. It is the same kind of inbreeding that can lead to an increased incidence of inherited diseases in human beings.

Another theory has it that inheriting different genes for a given characteristic produces superior expression of the characteristic, compared to inheriting two copies of the same gene, as happens with self-pollination. Neither theory explains all the evidence. And as Fitzgerald points out, some geneticists doubt the concept of heterosis altogether.

Nevertheless, by the early part of the twentieth century, many researchers believed that hybrid corn clearly had the potential to outperform traditional corn varieties. But production of hybrid seed was considerably more difficult and expensive than the production of conventional seed. Further progress was necessary before hybrid corn would replace traditional varieties. Again, that progress would contribute to the conditions that would lead to the Corn Belt epidemic.

The critical step was devised by the botanist Donald F. Jones in 1918, with the invention of the double-cross hybrid. In 1913, Jones began work in Connecticut on hybrids produced by crossing inbred lines of corn. Inbred lines of corn had predictable genetic characteristics, and so they were good for breeding experiments. The problem was that they did not produce much seed. That made them unsuitable for commercial seed production. Jones's idea was to produce hybrid varieties with increased yields and then cross those hybrids again. The experiment worked like this: Variety A and variety B were planted in one field to produce a hybrid designated AB. Variety C and variety D were planted in another field to produce a hybrid Jones called CD. These higher-yielding hybrids were then crossed with each other, producing a double-cross hybrid, designated ABCD.

When Jones began, the question was: Would the second cross increase the yield again? Or would it wipe out the gains made by the first cross? "Finally he planted the ABCD seed and compared the yields with those of standard corn," the National Academy recounted. "They were up by about 25 percent. Not by 2 or 3 percent, but 25 percent!"

Jones's discovery was a critical step toward the development of commercial hybrid corn, said Fitzgerald. "Even though the process was initially more cumbersome . . . the huge increase in seed produced was well worth the effort for commercial breeders," she wrote.[6]

One of the early proponents of hybrid corn was Henry A. Wallace, a university-trained plant breeder from a politically powerful farm family. His grandfather, Henry Wallace, founded a journal called *Wallace's Farmer.* He became prominent enough to be asked to serve on a presidential commission under Theodore Roosevelt.

Henry A. Wallace's father, Henry C. Wallace, was secretary of agriculture under president Warren G. Harding. Henry A. Wallace himself was secretary of agriculture under Roosevelt during the 1930s and vice president from 1941 to 1945. In 1948, Wallace went on to become the Progressive party's presidential candidate. Before he moved to Washington and entered politics, he founded the Hi-Bred Corn Company. It was the forerunner of Pioneer Hi-Bred International, now the nation's largest seed company.

Wallace was an early proponent of hybrid corn. He was quick to see the possibilities it offered and just as quick to see the difficulties in bringing it to market. Jones's work had solved the problem of producing large quantities of hybrid corn. But breeders still had much to learn about the breeding of hybrids. The results were often unpredictable. The best conventional varieties sometimes outperformed the hybrids, and some hybrid crosses were notable failures. The selection of appropriate parent strains for crossing was critical to the success of a hybrid offspring. But breeders were suffering through an extended period of trial and error.

The first commercial hybrid corn was introduced in 1929 by

Funk Brothers of Bloomington, Illinois. Others soon followed. In the 1930s, some farmers began trying the hybrids. Many followed customary practices that no longer applied to the use of hybrid seed. For example, they continued to select seed for the next year's planting, as they had done with open-pollinated varieties. But hybrid vigor persists for only one generation. Farmers who saved seeds were sadly disappointed when they saw that the second-generation plants the following year were smaller and less productive than the first generation had been. A crucial fact that fueled the expansion of the seed companies was that farmers had to buy new hybrid seeds each year. They could no longer save seeds.

Nevertheless, interest in hybrids continued to grow, partly driven by the proselytizing of the seed companies. By the end of World War II, hybrid corn had replaced traditional varieties. The private seed companies, including Funk Brothers, Pioneer, and others, said that the adoption of hybrids was due to the hybrids' superior characteristics.

Fitzgerald takes a revisionist view, arguing that other factors played a role, at least in the short term. One factor was unusually hot, dry weather in Illinois in 1934 and 1936. The weather reduced corn yield, including the yield of seed corn. With seeds in short supply, some farmers purchased hybrids rather than doing without. A second factor, Fitzgerald said, was an acreage-reduction program adopted by the Agricultural Adjustment Administration.

Farmers were paid to reduce their corn acreage. Some discovered they could switch to higher-yielding hybrids, reduce their acreage, collect the government subsidy, and still produce as much corn as before. One factor that did not drive the switch to hybrids was a need for more corn, Fitzgerald said. "Nobody was setting out to double the yield of corn," she said. "In the 1920s, farmers were burning corn for heat. There was way too much corn. The 1920s was the beginning of the farm depression. Suddenly they are left with all this corn, and there was no market for it."

Wallace understood what was happening, but he thought that government intervention could limit surpluses. "He was a big fan of government programs that would keep farmers from overproducing. He thought it was all right if more corn was produced, because farmers would just plant less," Fitzgerald said. Hybrid corn succeeded because "it was basically jammed through by plant geneticists."

Not everyone benefited from the advent of hybrid corn. Large seed companies were strengthened at the expense of smaller ones, and the benefits to farmers were mixed. "Hybrids bred to withstand specific adverse field conditions have made corn growing a more stable and predictable venture, but the social and economic costs have been considerable," Fitzgerald said.

"Not only must the farmers buy new seed each year, but hybrid corn introduced an array of corollary farm products such as fertilizer, insecticides, herbicides and other pesticides; the equipment used to apply these chemicals; and the enormous and enormously expensive machinery used to plant and harvest corn." The higher yield of the hybrids "has had the overall effect of sustaining chronic overproduction and declining farm prices."[7]

CRISIS IN THE CORNFIELDS

Although no one realized it at the time, the pieces that would contribute to the Corn Belt epidemic were falling into place. The transformation of the Corn Belt by hybrid corn was helping to create conditions that would make the epidemic possible. What was needed next was an infectious organism that could take advantage of those conditions. That crucial element was developing across the Pacific Ocean, ten thousand miles away.

The first hint of it came in 1962. That was when two corn breeders in the Philippines noticed an unusual problem in the local corn crop. A virulent new fungus called *Helminthosporium maydis* was attacking some hybrid corn varieties. The breeders collected data on the infection. They published their observations in an American breeders' newsletter. The infected Philippine corn had a certain resemblance to American corn, which might have raised questions about whether American corn was at risk. But it did not. American breeders who saw the report ignored it.

Three years later, the Philippine breeders collected more data and filed a second report. This time, they emphasized their concern about a possible infestation of American corn. Vulnerability to the fungus, they said, seemed to be related to a particular trait of the Philippine corn that was also found in most American corn.

This time, the report attracted the interest of Donald Duvick of Pioneer Hi-Bred International. He launched a small investigation. But he concluded that American corn was not susceptible to the *Helminthosporium* fungus. He blamed the infestation on something in the Philippine environment. Duvick was partly right. To his knowledge, American corn was not susceptible to the strains of *Helminthosporium.* What he did not know was that the Philippine outbreak was caused by a new strain. American hybrid corn would soon prove to be susceptible to the new strain. Duvick was one of few researchers to take the Philippine report seriously. Others continued to ignore it.

It was a costly oversight. In 1970, the new fungus, designated race T of *Helminthosporium maydis* (now called *Bipolaris maydis*), spread wildly through America's corn crop, especially in the Southeast. The disease it caused, southern corn-leaf blight, destroyed 15 percent of the U.S. corn crop. In some states, losses reached 50 percent. More than 1 billion bushels of corn were lost, costing farmers $1 billion.[8] The outbreak was one of the most serious epidemics ever to strike American agriculture.

In the 1960s, while breeders ignored the warnings from the Philippines, *Bipolaris maydis* was quietly gathering its strength. The first signs that the fungus could pose a problem in the United States appeared in 1968. "Out in the heartland, on a few isolated seed farms in Illinois and Iowa, a mysterious disease was producing 'ear rot' on corn plants," wrote Jack Doyle of the Environmental Policy Institute in Washington, D.C.

The losses that year were insignificant. The problem was written off as "a freak occurrence that would most likely die off over the winter, one of the 'normal' consequences of doing business with nature," Doyle said. In 1969, the phenomenon recurred in a few more states, including Florida. Again, the disease was dismissed as little more than a nuisance.

Bipolaris maydis did not reach epidemic proportions until 1970. The outbreak began in Florida. In May the fungus began attacking corn in southern Alabama and Mississippi. By June it had spread through southern Louisiana and parts of Texas. "The new fungus moved like wildfire through one corn field after another," Doyle wrote. "Within twenty-four hours it would start making tan, spindle-shaped lesions about an inch long on plant leaves, and in advanced form it would attack the stalk, ear shank, husk, kernels and cob." In the worst infections, "whole ears of corn would fall to the ground and crumble at the touch."[9]

The U.S. Department of Agriculture was "caught completely off guard," Doyle wrote. "On August 1, 1970, a time when millions of acres of corn in the Southeast had already been laid waste by blight, agriculture officials were confidently predicting a record 4.7 billion-bushel corn crop. A week later, they began revising their estimates downward." Reports of the blight led to turmoil in the corn market.

The fungus quickly spread northward to the Corn Belt of the Midwest. Although *Bipolaris* thrived best in the warm, humid conditions in the Southeast, it did some damage in the Corn Belt, too. In heavily infested fields, the crop was almost completely destroyed. "When severely diseased fields were picked, clouds of spores boiled blackly above the machines," the National Academy of Sciences wrote in its account of the epidemic. By September the blight had reached Minnesota and Wisconsin in the north and had spread westward as far as Oklahoma and Kansas.

Floyd and Ernest Brooks were young farmers in their twenties when the blight hit their adjoining farms about twenty-five miles southeast of Albany, Georgia, between the towns of Doerun and Sylvester. "I had around one hundred acres of corn," Floyd recalled. "That's usually what I had back then." Floyd and his brother had been farming since the early 1960s in the gently rolling countryside of southern Georgia.

"It's pretty good farming there," Floyd said. In 1970, the brothers were harvesting about seventy or eighty bushels of corn per acre. Floyd raised it to feed his hogs. Ernest had hogs and a few cattle. For most of the summer the corn seemed fine. The weather was good, as well as the brothers can remember, and the corn was thriving.

About midway between planting and harvest, they began to

suspect something was wrong. "It was in the part of the season when [the corn] starts silking. We noticed a discoloration in the corn," said Ernest. "The leaves began to turn a little rusty color, right on up to the top. That's unusual. You don't see that unless something's wrong."

At first he didn't know what was wrong. "We'd never had any experience with anything like this. We knew we had some problems, because we'd never seen corn do this discolor from bottom to top." By the time Ernest and Floyd realized they were in trouble, they had gotten word from the agricultural extension service that a new blight was moving in.

"We didn't make any corn that year," Floyd said. Once the blight had struck, nothing could be done to stop it. "The corn started going down pretty quick. It started losing its color. The plant itself started turning back, it started turning back yellow. It was pretty much the whole field." That happened to "everybody around that had the varieties that weren't resistant."

Floyd was particularly distressed by the drought, because he had made a change that year. "They had some kind of campaign," he said, referring to a suggestion that farmers should plant two new corn varieties. "You plant two rows of this and two rows of the other variety. It was supposed to improve our yields by mixing them two and two. But as it turned out, they both got the blight and didn't do very well."

As the blight spread, the extension service realized that one particular variety grown in the area was resistant to the blight. Floyd wasn't terribly happy to hear that piece of news. "I had been planting it up until that year, and that year I changed." If Floyd hadn't adopted the new varieties, he would have been spared.

The hogs still had to be fed whether Floyd made corn or not. "We had to turn around and buy corn," he said. "The best I remember, they shipped some corn in. We didn't have any problem getting the corn." But the price tripled, from about one dollar per bushel to three dollars. "I think it was the first time we'd ever heard of three-dollar corn," he said.

Along with a couple of drought years that hurt the corn and caused the peanuts to fail, Ernest and Floyd remember the corn blight as one of the worst disasters they have experienced. "It just almost wiped the corn out," said Ernest. The strength of the hog market helped them survive. "We made it through all right," said Ernest. "You just sort of grin and try to bear it," said his brother.

The Philippine corn breeders who sounded the alarm about the epidemic noted that Philippine corn varieties and most American corn varieties had something in common: a single gene that made the corn plants sterile. They produced no pollen. The gene altered the complex chemical balance in the plants' cytoplasm, the sticky liquid inside individual cells. The gene produced what was called "male-sterile cytoplasm." It was bred into various American corn varieties because it simplified the production of hybrid corn.

Hybrid corn is produced by fertilizing one line of corn with pollen from another line. Before researchers discovered male-sterile cytoplasm, the tassels at the top of each corn plant—where the pollen is formed—had to be laboriously clipped off each of thousands of corn plants to prevent them from making pollen.

The Philippine breeders noted that the fungus infected only those corn plants that had the male-sterile cytoplasm.

The cytoplasm was known as Texas cytoplasm or T cytoplasm, because it originated in a Texas corn plant. In 1970, 85 percent of U.S. corn acreage carried T cytoplasm.[10]

"The corn crop fell victim to the epidemic because of a quirk in the technology," the National Academy of Sciences said later. "In one sense, they [the corn plants] had become as alike as identical twins. Whatever made one plant susceptible made them all susceptible."[11] If breeders had recognized that, they could have developed varieties without that gene, and much of the crop could have been saved.

The epidemic was abetted by ideal weather in the spring and summer of 1970—ideal weather for the fungus, that is. *Bipolaris* spores are carried by the wind to adjacent corn plants. There they require moisture to germinate and infect the healthy leaf. Once the spores germinate, they invade the leaf, reproduce, and produce new spores.

If the weather is too hot or too cold, this process slows down. During most of 1970, the weather was just right. Sufficient rain fell to allow the spores to germinate. Temperatures were suitable for speedy reproduction. In the northeastern states and in the western portion of the Corn Belt, a shift to dry weather in September saved much of the corn crop. Without that shift, the losses could have been much higher than the estimated 1 billion bushels.

"In other words," wrote Doyle, "the nation was lucky."

By the time the academy's report was issued in 1972, breeders and seed companies had replaced the susceptible corn plants. Emergency efforts to acquire new seed and to breed new hybrid varieties that did not carry the male-sterility gene headed off further crop losses.

In most other respects, however, the corn crop remained genetically uniform. So did other major crops around the world. In 1848, for example, French wine grapes were struck by a disease called powdery mildew. The outbreak was controlled in the short term by applications of sulfur dust. To put the disease to rest over the long term, the French imported American grapes, which were immune to powdery mildew.[12]

But the grapes brought with them the aphid *Phylloxera vitifoliae.* It did more damage than powdery mildew. American grapes resistant to *Phylloxera* were then imported. They, in turn, carried the downy mildew fungus. That caused a third wave of disease in French vineyards. And the story is not over. In 1980, John Baritelle, a vineyard owner in California's Napa Valley, the leading wine-growing region in the United States, found four stunted vines in one of his fields. The following year, sixteen vines died.

Routine tests, intended to discover the cause of the deaths, led to a surprising and potentially catastrophic discovery: The genetically uniform rootstock used in nearly all of Napa's grapes was susceptible to *Phylloxera.* Eventually, most of the valley's grapes would probably succumb. Winegrowers were forced to rip out tens of thousands of grapevines, knowing they would eventually succumb. The costs of replacing them could climb into the millions.[13]

In 1916, an outbreak of red rust cut so deeply into the American wheat crop that the Hoover administration asked Americans to observe wheatless days every week to cut demand. In 1954, wheat-stem rust destroyed 75 percent of the wheat crop. In the 1920s, an epidemic of citrus canker required the destruction of 20 million citrus seedlings and trees in Florida.

Not all epidemics are due to uniformity of the affected crop. The French grape epidemics probably were. The wheat-rust epidemics probably were, too, according to the National Academy of Sciences. Citrus canker is a continuing concern in Florida, for only a few varieties of citrus are grown there. "As resistant varieties become available, farmers spread them over very wide areas. When the fungus mutates to a form that attacks the new variety, an epidemic results," the academy said.

Dutch elm disease, on the other hand, was not a consequence of genetic uniformity. It wiped out elms across the United States because they had little natural resistance to it no matter what their genetic makeup. (The disease was brought to the United States by American veneer factories, which took advantage of bargain-basement prices on diseased logs from elm trees in Holland. The logs "scattered the disease carrying beetles along the railroad rights-of-way out to the furniture factories in the Midwest," the academy said. "And the graceful elms that once lined the streets of America were lost in exchange for a few dollars' income for veneermen.")[14]

After the epidemic of southern corn-leaf blight, America's fields remained as vulnerable to an epidemic as they had been before. They were no longer vulnerable to the fungus. But with regard to any other diseases that might appear, nothing had changed. The stage was set for another calamity.

Garrison Wilkes is one of the very few scientists who understands what is at stake. "Never before have there been such widespread monocultures—dense, uniform stands of billions of plants covering thousands of acres—all genetically similar," he said.

Wilkes grew up on an apricot ranch in the 1930s and 1940s in the San Fernando Valley, when the valley was covered with ranches and farms. His father's ranch was in what is now North Hollywood, and there are two hundred half-million-dollar houses on it. The population of the valley has grown from eighty thousand, when Wilkes was a boy, to 2 million. Little trace of the ranches and farms remains.

A broad-shouldered, soft-spoken man with a philosophical bent, Wilkes is an expert on teosinte who in recent years has concentrated his efforts on calling attention to the problem of genetic uniformity. Although he grew up on a ranch, Wilkes did not set out to study agriculture. He was more interested in the kind of biology that might land him a job as a researcher with the National Park Service. But a course on the origin of cultivated plants changed his mind.

"That was exciting," he recalled, leaning back in his chair in his office at the University of Massachusetts. He applied to Harvard University for graduate school so he could study with Paul Mangelsdorf, the corn warrior.

"This was a day and an age when you could do things on a shoestring, without big grants," Wilkes said. Mangelsdorf bought Wilkes a used Land Rover for $1,200 and gave him $800 for gas. Wilkes had $3,000 in scholarship money. That added up to $5,000—enough for Wilkes to spend a year in Mexico studying teosinte. He traveled fifty-six thousand miles, crisscrossing hundreds of Mexican farms and fields.

"Teosinte is the closest wild relative of maize, and nobody had ever found all the populations and cobbled this all together. It seems unbelievable that a plant that important had been left," Wilkes said.

That was in the 1960s. On more recent trips to Mexico and elsewhere in Central America, Wilkes has been distressed by the rate at which teosinte is disappearing. For Wilkes, that is an indication of change for the worse.

"I saw in the 1960s, in Mexico, something that wasn't terribly different than the California I grew up in, in the 1930s and 1940s. Mexico in 1960 was like California in 1900. There was that openness, that can-do attitude. I was treated very nicely. Farmers talked to me. Now it's gotten a lot like here. They say, 'What's in it for you? Why are you really doing this?' "

The shrinking of the population of wild teosinte led Wilkes to concentrate his energy on the problem of disappearing crop germplasm. "Food security and biodiversity will be our most obvious challenges in the decade of the 1990s," he said. "How can we supply adequate nutrition to the 5.5 billion humans that exist now and the more than 6 billion that will exist by the year 2000?"

Wilkes used a little arithmetic to dramatize the problem. In the first twenty years of the next century, human beings will need to produce as much food as they have since the beginning of agriculture ten thousand years ago.

"Given the needs of the future, genetic resources can be reckoned among society's most valuable raw material. Any reduction in the diversity of resources narrows society's scope to respond to new problems and opportunities."

The pace of finding and evaluating new germplasm has already fallen off. It will not be apparent until the end of the 1990s because varieties being released for commercial use now are the result of work begun ten years ago, Wilkes said.

For decades, American agriculture has been based on the premise that uniformity is good. Supermarket bins overflow with identically tinted, uniformly sized oranges. Ears of corn, all the same color and length, are stacked in neat pyramids. Plastic bags hold radishes that all glow with the same ruby skin.

Farmers are served by the uniformity, too. It is no accident that each plant in a cornfield is exactly the same height as its neighbors. Breeders have installed that uniformity in the corn to facilitate mechanical harvesting. Height is one of many aspects of commercial crops that is regulated by breeding.

Farmers want all the seeds in their fields to germinate, flower, and mature at the same time. They demand predictability. They want to know that the produce will be uniform in terms of nutrition, taste, texture, and so on. It is all part of what the crop scientist Jack Harlan called the "pure-line mentality."

The danger is greater in regions such as the American grain belt, where a uniform climate allows similar or identical crops to be grown across hundreds of miles. Cultivated plants entirely replace natural flora, providing unrestricted entry to such organisms as *Bipolaris maydis.*

The defense is to maintain "patches and corridors of other

crops, as well as natural vegetation, to slow down the spread of a pest or disease," said Gary Paul Nabhan. Without such protection, he said, "insects or pathogens can rapidly multiply and devastate large areas of one crop."

AFTERMATH OF THE EPIDEMIC

In August 1970, as the extent of the epidemic of southern corn-leaf blight became clear, seed producers began to worry about the following year's seed supply. Much of the nation's corn seed was produced in the midwestern Corn Belt, where the blight was hitting. In the spring of 1971, seed producers began selling susceptible seed to farmers in western and northern states where the weather was not favorable to the blight. Some farmers received a mixture of half-susceptible seed and half-resistant seed, according to Doyle. No one knew quite what to expect in 1971. Would the blight strike again? Would it be worse than it had been in 1970?

As luck had it, the weather in 1971 conspired with seed producers and farmers to help defeat the blight. The weather was cooler and drier. Losses to southern corn-leaf blight were much lighter. The crisis had passed.

Breeders quickly began to breed normal cytoplasm into the sterile varieties. During the winter of 1970–71, seed from normal cytoplasm was produced in Mexico, Hawaii, and Argentina. By 1972, seed producers were supplying farmers with blight-resistant seed.

"The key lesson of 1970 is that genetic uniformity is the basis of vulnerability to epidemics," the National Academy of Sciences said. The meat of the report was an assessment of the genetic uniformity of American crops. More than twenty years after its completion, it remains the most thorough examination of the subject.

The academy's findings were stunning. They found that a mere six varieties of corn accounted for 71 percent of the American corn crop. The story was the same with other crops. Nine varieties of wheat made up half of the wheat crop. Four potato varieties accounted for 72 percent of American potatoes. Two types of peas produced 96 percent of the pea crop. A single variety of sweet potato, the Centennial, accounted for 69 percent of the American harvest.

"The striking uniformity among American crops is no accident," the academy reported. Consumers, food processors, and the designers of farm machinery all put pressure on farmers to produce uniform crops. That forced breeders to abandon genetic diversity and to rely on pest and disease control experts to protect the crops. "That the nation has experienced so few epidemics provides convincing evidence that their efforts have been remarkably effective," the academy concluded.

The academy determined that the adoption of hybrid corn had seriously reduced genetic diversity throughout the Corn Belt. Breeders increasingly focused on improving existing inbred lines rather than searching for new varieties that could add genetic diversity to commercial seeds. The reason was simple: Developing new inbred lines from open-pollinated varieties takes longer than improving existing inbred lines.

The use of landraces or wild relatives of corn, which could vastly expand the crop's genetic diversity, is the most difficult and time-consuming process. It can take decades to transfer a disease-resistance gene from a wild-crop relative into a commercial variety. The academy noted that sizable collections of corn germplasm are available in seed banks. But the availability of the collections doesn't ensure that they will be used.

"Exotic materials tend to be poorly adapted and therefore low yielding under conditions prevailing in the United States," the academy said. "In addition, they usually exhibit other plant or ear characteristics that make them unacceptable to farmer or consumer."

Consumers demand the best food at the lowest cost. To provide that, farmers must use the highest-yielding varieties that are resistant to insects and disease—and adaptable to mechanical sowing and harvesting. The crops must also be able to withstand shipping and processing. Only a limited number of varieties have been developed to meet those stringent requirements.

The demands for uniform crops have been strong enough to undermine any attempt to broaden the genetic diversity of American crops. The academy's warnings were strong and clear. But the economic imperatives were stronger. Seed producers, farmers, food processors, and consumers simply were not going to accept what they saw as inferior varieties. The threat of famine—the collapse of a crop in the face of an epidemic—seemed too remote.

The academy made several recommendations. It urged breeders to conserve genetic diversity in wild plants and landraces. Explorers could "go back to the geographic area where the crop seems to have originated in a search for useful genes from the wild types occurring there or even from varieties that local farmers grow," the academy said.

The academy's report launched a minor industry in the preparation of research reports reaffirming the problem of genetic uniformity. A year after it was released, an Agriculture Department study group called for more research on the utilization of exotic germplasm in American crops. The U.S. effort to preserve germplasm "has been too haphazard, unsystematic and uncoordinated, and has never received the high priority it deserves among the many agricultural research programs," the study group said. "The situation is serious, potentially dangerous to the welfare of the nation, and appears to be getting worse rather than better."[15]

The problem has been pointed out again and again in study after study by the academy, the General Accounting Office, the Office of Technology Assessment (OTA), and the Agriculture Department. Still, almost nothing has been done to correct it. If the effort that was put into all of the studies had been directed toward the expansion of crop diversity, the problem might have been solved long ago.

As studies were spinning off the printing presses in Washington, Donald Duvick of Pioneer began his own investigation of the uniformity of U.S. crops. Duvick, who rose to become vice president of research at Pioneer before retiring, is a respected crop geneticist and researcher. Although neither he nor Pioneer were singled out in the series of reports decrying the nation's genetic vulnerability, Pioneer, as the nation's largest seed company, could be considered partly responsible for the nation's genetic vulnerability.

At a seed-industry conference in 1981, Duvick challenged the assertion that plant breeders were responsible: "The reason for concentration on a few varieties is said to be that plant breeders release very few new varieties and, further, that they develop new varieties only from crosses of a few of the highest yielding older varieties," Duvick said, recapping the argument. "In these ways, it is said, breeders successively eliminate most of the diverse kinds of germplasm that theoretically now might be growing on U.S. farms. . . ."

According to Duvick's investigation, however, breeders were not eliminating biodiversity. He surveyed breeders working on corn, soybeans, wheat, sorghum, and cotton. The corn breeders told him that 212 hybrid varieties were close to commercial release. Some seventy-four hundred more were in the experimental stage, and research was beginning on another sixty-one thousand varieties. "Our backup reserves of hybrids, inbreds and breeding populations are extremely numerous," Duvick concluded.

There was a problem with that analysis, however. Although it seemed to show that breeders were working with an enormous amount of genetic diversity, many of the thousands of varieties under study were probably very closely related to each other.

Duvick has conceded that that is probably true. But it has been difficult to confirm the genetic similarities in different varieties. The reason is that seed producers are unwilling to disclose the pedigrees of their seeds to competitors. Stephen Smith, the Pioneer corn geneticist, devised a way to resolve that question. He came up with a method of assessing genetic diversity in commercial corn varieties that does not require disclosure of confidential pedigrees. It is a genetic analysis that lets him determine how closely related two varieties are without disclosing what germplasm went into them.

Smith's analysis confirmed the worst suspicions. He examined 138 commercial hybrid corn varieties used in the United States and determined that 56 percent of them were genetically distinct from one another. But the remainder—nearly half of the varieties Smith examined—fell into only seven genetic groups. Within each group, the different varieties were nearly identical. U.S. corn crops were heavily dependent on only four inbred lines, Smith found. Farmers who try to reduce their risks by planting different varieties may be planting essentially the same thing, under different names. They have no way of knowing, Smith said in his report.

The solution will not be simple, even if seed companies are persuaded to do something about the problem. Incorporating new genetic material into commercial crops is a difficult and lengthy process, as James Quick's experience demonstrates.

"I don't think we can afford to become complacent," Smith said. "There's a lot of genetic diversity" in corn breeding lines. "But plant breeding is such a long-term proposition . . . that once you get it in a bloody mess, you've really got trouble sorting it out."

Nevertheless, he said, "you've got to think long-term. The problem is that economic incentives all seem to be for the short term. And that's true for any industry."

Duvick's survey confirmed that closely related varieties can cover huge areas—precisely the problem the National Academy of Sciences and all the other report writers had warned about. In 1970, the single inbred corn line, designated B37, for example, was used in hybrids covering 26 percent of U.S. corn acreage in 1970. By 1979, hybrids containing B37 genes covered only 2 percent of U.S. corn acreage, the breeders reported. Those varieties had exhausted their useful commercial lifetimes.

Duvick concluded his investigation by asking breeders whether the problem of genetic vulnerability in corn was serious. Only one told Duvick it was a problem. Sixty-four percent said there was nothing to worry about.

LESSONS UNLEARNED

More than two decades after the academy's report, the lack of genetic diversity in American crops remains a serious problem.

Sunflowers, for example, are a $200-million-a-year crop. Ninety percent of the hybrids grown in the United States share the same cytoplasm. If a pest or a disease finds a chink in that cytoplasm, virtually the entire crop could be wiped out.[16]

Domesticated chili peppers contain only a small fraction of

the genetic variability that exists in their wild relatives. Meanwhile, deforestation in the tropics is wiping out many populations of chili peppers, destroying diversity that could potentially be bred into cultivated varieties.

Furthermore, collections stored improperly in gene banks are being lost. Chili-pepper seeds stored in a Guatemala gene bank, for example, are nearly dead. If the collection is not rejuvenated immediately by planting and multiplying the seeds stored there, the samples in the collection will become extinct.[17]

Potatoes are a particularly good example of the continued genetic uniformity of American crops. The U.S. potato crop was worth $2.5 billion in 1989, making it one of the nation's most important crops. Yet the entire crop consists of only six varieties.[18] The principal variety is the Russet Burbank, the long, brown potato commonly associated with Idaho.

The Russet Burbank has a higher ratio of solids to water, which makes it ideal for french fries. "It was bred very unscientifically, but it's got most of the market," said John Bamberg, curator of the Agriculture Department's Interregional Potato Introduction Project in Sturgeon Bay, Wisconsin.

It is the potato of choice for McDonald's french fries. Eighty percent of McDonald's Russet Burbanks come from J. R. Simplot of Idaho, who signed an agreement with McDonald's in the 1960s and rode its coattails to become one of the nation's wealthiest men. Before Simplot and McDonald's forged their alliance, about a thousand farmers in eastern Washington State grew different varieties of potatoes on about twenty thousand acres. By 1994, half as many farmers had planted mostly Russet Burbanks on 115,000 acres.[19]

As McDonald's has expanded around the world, it has taken the Russet Burbank with it, so that more and more of the world's potato acreage is now planted with this single variety. "What appears [to be] a genetic godsend and economic bonanza for that company today could well become an economic nightmare for them tomorrow . . . should one tiny organism find a genetic window of virulence in the Russet Burbank potato," wrote Jack Doyle. If that happens, he said, McDonald's "will have contributed mightily to the spread of a genetic epidemic. And for that, they would be culpable, and perhaps should be liable."[20]

In 1990, it happened. Agriculture Department researchers turned up a new variety of *Phytophthora infestans* in Vancouver, British Columbia, and Athens, Pennsylvania. Potato blight had returned with a vengeance. "The disease is remarkably explosive," said William Fry, a plant pathologist at Cornell University. "An affected field looks like it has been burned."

Potato blight is once again ravaging potatoes, including the Russet Burbank. What is worse, some of the new variants of the fungus are resistant to metalaxyl, the fungicide most commonly used to control *Phytophthora.* "The disease is going to be more difficult to control than it used to be," Fry said. "How much more difficult, we're going to have to wait and see." The new *Phytophthora* variety, which probably originated in the central Mexican highlands, is quickly spreading around the world. Americans who dine at McDonald's may have to wean themselves from the Russet Burbank.[21]

In 1994, some farmers in upstate New York reported that as much as half of their potato harvest had succumbed to the blight. In Maine, the potato crop was down by 13 percent, and it declined by 18 percent in Pennsylvania. Vyron Chapman, who has five hundred acres in Cassville, New York, and who usually produces 13 million pounds of potatoes, had losses worth $500,000. "It's about time to buy seed potatoes for next year, but I'm not even making an attempt yet," he said. The problem was finding disease-free potatoes.[22]

Walter Stevenson, a plant pathologist at the University of Wisconsin, said the blight appeared in Wisconsin potato fields in July, and it continued to spread until harvesttime. In some fields, Stevenson saw evidence of infection in 20 percent of the crop. "That's a huge amount," he said. "In most years, one percent would be high." The blight came into Wisconsin with seed potatoes, which the state imports because it can't produce enough seed of its own. Wet weather and warm temperatures late in the growing season encouraged the infection to spread, Stevenson said.

Even after harvest, the blight continued to spread. In Wisconsin, potato growers store their harvest in huge warehouses, where a single pile of potatoes can be twenty feet high and cover an area the size of a football field. In September, in many of those warehouses, the blight was slowly turning potatoes into a putrid, black mushy syrup. As each rotted, it dripped on others, spreading the infection, until a large percentage of the potatoes had turned to mush. "There was one warehouse of potatoes that were harvested in the middle of September," Stevenson said. The twenty-foot-high pile of potatoes dropped five feet "in a matter of days," Stevenson said, as the potatoes turned to mush and the black syrup drained away.

Despite the extent of the losses, the problem failed to gain national attention. Because of bountiful harvests in other states, the national potato crop in 1994 was up 7 percent over that of the previous year. The problems in New York, Wisconsin, Maine, and a few other states were easy to ignore.

A report like the 1972 National Academy of Sciences report can never be done again. The Agriculture Department not only did not take action to correct the problem pointed out in that report, it stopped monitoring the amount of land devoted to certain crop varieties. Thus, the department made it impossible for the problem even to be monitored.

In 1994, the academy tried to update the 1972 report, with help from Garrison Wilkes. In almost every case, the situation had become worse.[23]

In 1972, for example, the academy noted that two varieties of beans—the Michigan navy bean and the pinto bean—accounted for 60 percent of the 1.5 million acres of beans in the United States. The report noted that dry states, in particular, were especially susceptible to an epidemic.

In 1981 and 1982, the prediction came true. Colorado and Wyoming, the two states at greatest risk, had an epidemic of rust that caused $15–$20 million in damages. From 25 percent to 50 percent of the pinto-bean crop was lost, "causing the widespread destruction clearly recognized as a potential in dryland regions in the 1972 report," the academy said.

A milder rust epidemic struck the region again in 1987. The

Agriculture Department began scanning the seed banks to find beans resistant to rust, and it developed several resistant breeding lines. But susceptible varieties of pinto beans still make up 40 percent of the crop in Colorado, southwestern Nebraska, and northwestern Kansas.

Four hundred varieties of winter wheat used in the United States can all be traced to seventy-four ancestors. Most of the germplasm comes from a small number of varieties used before 1919. Wheat breeders are beginning to remedy that situation, but they have a long way to go.

The six corn varieties that had accounted for 71 percent of the American crop at the time of the 1972 report made up 43 percent of the crop in 1980. Breeders had incorporated almost none of the germplasm available from wild-corn relatives like teosinte. Less than 1 percent of the germplasm in corn came from something other than North American breeding lines.

"The current haphazard, uncoordinated, and unsystematic approaches to the problem reflect dangerous and inappropriate national and global priorities," Wilkes said in 1992 at the annual meeting of the American Association for the Advancement of Science. "It is difficult to visualize a challenge more profound in its implications yet less appreciated by the general public."

To help the academy assess genetic vulnerability, Wilkes collected and examined reports from what are called crop advisory committees, established by the Agriculture Department. These committees, made up of volunteers, are each charged with providing guidance and policy recommendations for a single crop. Because they are the work of volunteers, the reports appear irregularly and vary in completeness.

It is not clear who reads them or whether their recommendations are taken seriously in the Agriculture Department. But Wilkes found some of them full of information about crop uniformity. He looked at reports on twenty-six crops, ranging from corn and wheat to such minor crops as pecans, cucumbers, and muskmelons.

Eleven of the twenty-six crop advisory committees said that they found enough genetic uniformity to make an epidemic possible, with some saying the situation was catastrophic. Apples, peas, pears, sugar beets, sugarcane, sweet potatoes, cucumbers, and grapes were all reported to be at serious risk of epidemics.

"Most minor crops, at least within the U.S., have extremely narrow genetic bases and are already susceptible to numerous diseases and pests," Wilkes said. "The fruit crops are particularly vulnerable to epidemics in the near future, because the chemicals which have buffered them from disaster over the last three decades are now being withdrawn from the market."[24]

"I still feel that the genetic vulnerability issue is one where we have no national strategy, except that the more breeders, the more chance to have diversity incorporated into the varieties," said Henry L. Shands, director of the National Plant Germplasm System. "We need more breeders infusing diversity into the mix."[25]

The broad reach of modern agriculture now means that genetic uniformity can spread across millions of acres, across a nation, or across a continent. The Irish potato famine and the 1970 epidemic of southern corn-leaf blight, said Wilkes, are "early warnings of the dangers."[26]

Chapter 5

One Planet, One Experiment

On a balcony in a dimly lit back corner of Harvard University's Museum of Comparative Zoology stands a narrow glass case. It is part of an old, dusty exhibit of birds of the world. On the bottom shelf, where visitors must stoop to see it, is a large, mostly black bird with red feathers on the back of its head, bits of white on its wings, and yellow eyes. It is a stuffed specimen of the male imperial woodpecker, *Campephilus imperialis.* At twenty-two inches from head to tail, it is the world's largest woodpecker.

The museum's specimen was collected in the Mexican state of Chihuahua on September 8, 1905, during an expedition organized by John E. Thayer. A Harvard graduate and bird fancier, Thayer was described by a friend as "the finest example of what a New England country gentleman ought to be." He spent a lifetime collecting birds, and in 1932 he donated twenty-eight thousand of them to the Harvard museum, including the imperial woodpecker.

Few visitors see the woodpecker. It is not listed in the museum's catalog. It was misnumbered and apparently unrecorded in Thayer's own notebooks. The specimen is of little scientific value;

ornithologists studying woodpeckers are more interested in the fifteen other specimens locked away in the scientific collection several flights up from the exhibit area. But that lone imperial woodpecker has special significance for Harvard biology professor Edward O. Wilson, who passes it daily on the way to and from his office.

"That bird has been a talisman for me," Wilson said. "It became extinct thirty years ago. One of the last two known individuals was shot and eaten by a Mexican truck driver." When the truck driver was later asked about the bird, he said, "It was a great piece of meat."

The bird reminds Wilson, he said, "that we are not masters of this world." Or, "We're masters in the sense we could destroy it."

Wilson has been at Harvard since 1950, first as a graduate student and later as a professor, but his speech retains the rhythms and inflections of Birmingham, Alabama, where he was born in 1929. He is slim and long-limbed, with sharp features and the flinty look of a New Englander despite his southern origin. He wears the kind of tweed jackets that one expects to see on a Harvard professor, and his enthusiasm for his research can cause him to erupt in a childlike burst of laughter when he talks about it.

Wilson has become one of the most powerful voices among those calling attention to the global loss of biological diversity. He has no special interest or expertise in agriculture. He is camped among the environmentalists who warn about the loss of the tropical rain forests and the threats to the spotted owl.

His effort to assess the rate of species extinctions around the world provides a broader context for what is happening with agricultural resources. The disappearance of natural habitats takes with it not only mountain gorillas, pandas, and imperial woodpeckers. It is also responsible for the disappearance of countless wild relatives of important food crops.

Wilson is an accomplished writer. Two of his thirteen books have won Pulitzer Prizes. He won the National Medal of Science in 1977 and the Crafoord Prize of the Royal Swedish Academy of Sciences in 1990. He is the recipient of ten honorary doctorates. His office, in a modern building adjacent to the Harvard museum, is like scores of other scientists' offices. It has a desk, bookcases, several long tables for experiments, and a few pieces of laboratory glassware scattered about. It takes a moment or two to notice, but there is something about Wilson's office that distinguishes it from the offices of his Harvard colleagues. On one side of the room, a countertop is alive with hundreds of crawling ants.

Amid microscopes and books, leaf-cutter ants stream in endless columns along a three-foot twig looping from one plastic box to another beside Wilson's desk. Each ant carries a tiny, snipped piece of oatmeal. "These are leaf cutters, a dominant species of the tropics, from Texas to Patagonia," Wilson said proudly. In the summer, he periodically throws a handful of leaves into the ants' box. In the winter, when he cannot get fresh leaves, he feeds them rolled oats.

The ants have a sophisticated social system. They live not off the leaves but off a fungus that grows on them. The ants are carrying the pieces of oatmeal to a fungus "garden" where the leaves will be delivered as "food" for the fungus.

Leaf cutters harvest leaves from crops as well as other plants. An attack by leaf-cutting ants can have serious economic consequences for a farmer in the tropics. A single nest can cover more than six thousand square feet, and a single queen can have 150 million daughters.

Like other animals, insects, and plants, the leaf cutters have a valuable role to play in the forest. They prune plants, aid the breakdown of plant material, stimulate plant growth, and turn the soil. "Leaf-cutting ants are among the most advanced of all the social insects," Wilson said.

Wilson has spent his life as a myrmecologist—a student of ants. He is now engaged in a reclassification of six hundred species in the *Pheidole* genus, a group of New World ants. It is a Herculean task that even fellow myrmecologists said he was crazy to attempt. For that project, he is dipping into Harvard's ant collection. Numbering about 1 million specimens, it is the largest ant collection in the world. Of the six hundred *Pheidole* species in the collection, more than half are new to science. Wilson is laboriously examining each specimen, noting its features with great precision, and using those notes to establish criteria distinguishing one species from another.

Wilson carefully placed a preserved specimen of *Pheidole longiscapa* under a microscope, adjusted the focus, and offered a visitor a look. The strikingly beautiful image was of a black-eyed monster with a gleaming golden head and long, sweeping antennae. It stared directly back into the microscope.

"I call it looking in the face of creation," Wilson said. "You're looking at something that may be a million years old, and nobody's seen it before."

Wilson summed up much of his work on ants in a 1990 book entitled, simply, *The Ants*. Weighing in at about 7.5 pounds—or one hundred thousand times the weight of the typical ant—the book led to the creation of the computer game SimAnt, and it won Wilson his second Pulitzer Prize.

Few scholarly pursuits could seem more removed from the urgency of the biodiversity crisis. Yet Wilson's pursuits have thrust him into the center of angry public debates. He has become one of the most controversial and important figures in modern biology.

In the 1970s, Wilson was hounded by demonstrators and bitterly denounced as a racist. The issue was his creation of a new field called sociobiology. In a book of the same name, he argued that biology was important in shaping social structure and relationships. Critics said he was providing justification for an inequitable status quo. They said he was promoting genetic determinism, a social survival of the fittest.

Wilson said he was merely trying to apply evolutionary biology to human behavior. He noted, for example, that male dominance is common in many animal societies. That probably means it has a biological origin. His conclusion was that "men are likely to continue to play a disproportionate role in political life, business, and science." The reaction was predictable. "The activists wanted to brand this as not only wrong but sinful," he said.

Wilson was then a tenured Harvard professor and already a prominent figure in biology. But the depth of the protests frightened him. "I thought my career was going up in flames," he said.

"In time I gave up open public lecturing, confining myself to talks and seminars to universities, colleges and professional groups," he wrote in an autobiographical sketch a few years ago.

He was particularly disturbed that some of his severest critics were his Harvard colleagues, including the equally well known biologist Stephen Jay Gould. In time, the furor eased. Wilson's career did not go up in flames. Although he continues to have critics, much of his work on sociobiology has become part of mainstream biological thinking. In 1989, the Animal Behavior Society recognized his book *Sociobiology: The New Synthesis* as the most influential book in the field in twenty years.

Early in the 1980s, Wilson felt he had completed the work on sociobiology, and he turned to research on biological diversity. What Wilson has attempted to do, as thoroughly as it has ever been done, is to estimate the rate at which the earth is losing species.

Wilson has what he calls his "sound bite" version of the biological-diversity crisis, which he explored in his 1992 book *The Diversity of Life.* He related it at the Harvard Faculty Club over a dinner of hare consommé, wild boar, mallard duck, and blackened venison. ("I hope you don't say we feasted on a medley of endangered species," Wilson said, chuckling, as one course followed another. "We still haven't hit any endangered species.")

"The diversity of life on earth is far greater than even most biologists recognize," Wilson said. Fewer than 10 percent of the earth's species have scientific names, making earth "a still mostly unexplored planet."

Wilson is a systematist, in the tradition of Charles Darwin and Carolus Linnaeus. "The great figures of biology were systematists," he said. "This was big-time biology. They went on expeditions and collected specimens," and they spent their lives categorizing those specimens and trying to derive biological principles from them. It is the same thing Wilson has done with ants.

At the turn of the century, however, that kind of exploration of the earth began to diminish. Biology became preoccupied with the study of cells and the basic processes operating in living organisms. By the 1970s, cellular biology had become molecular biology as researchers probed the workings of the genes inside the cells. Systematics had become a forgotten backwater in biology, peopled by aging scholars unable to attract students.

"Patronage is all-important in science," Wilson said. "The molecular biologists made great discoveries because they were rich, well endowed. They did not become rich because they made great discoveries. Patronage did not exist in biodiversity and systematics. If you had support, the brightest students would be pouring into the field—if they knew they could get jobs."

That is exactly what happened with the rise of the environmental movement. "A new mission was provided for systematics," Wilson said. "It became clear that the great effort to describe life on earth was not completed. In fact, in a statistical sense it has barely begun. We understand the forces that hold ecosystems together less than we understand what holds atomic particles together."

It would not take a lot of money to begin to remedy that, Wilson said. This is what he calls "little big science," compared with, say, the human-genome project. "A lot could be done for one hundred million dollars."

It is only a few steps from Wilson's office to Harvard Yard, which has been trod and explored by Harvard students for hundreds of years. But for Wilson, even something as familiar as Harvard Yard is a trove of unexplored biological riches. "No one could guess the number of bacteria in Harvard Yard, but conservatively it would be in the tens of thousands," he said. "Harvard students could be studying previously unknown bacteria in the ground right outside their dormitory.

"That diversity, on a still mostly unexplored planet, is a potential source of immense wealth and benefit for people—in the form of medicine, new foods, petroleum substitutes, and other amenities. It provides absolutely vital ecosystem services we take for granted, little things like clean air and water and the fertility of the soil. It's also the source of almost unlimited scientific information in the future and still unfathomed aesthetic and spiritual enjoyment. Taken country by country," Wilson said, "it's part of the national heritage."

Wilson stopped. "The sound bite has gone on too long," he said. "Ted Koppel would have already cut me off."

MASS EXTINCTIONS

No one can say with certainty how much genetic diversity has been lost or how fast it is disappearing. Many biologists say that the world is now in the midst of a catastrophic extinction episode rivaling the great mass extinctions of prehistoric times, such as the one that led to the demise of the dinosaurs. All of the others occurred millions of years before human beings overran the earth. The tragedy of the current episode is that humans are here not only to experience it, they're also responsible.

For Niles Eldredge, a paleontologist at the American Museum of Natural History in New York, the disappearance of many of the migrating songbirds that once filled temperate skies every spring is a sign of trouble. "In the days before sophisticated sensing technology, miners often took canaries with them as they ventured down into the bowels of the earth," Eldredge wrote in *The Miner's Canary,* a study of mass extinctions.[1] Noxious gases would fell the canaries, alerting the miners to the danger before they were overcome themselves.

For the first time in history, humans now are able to affect the earth's environment on a global scale. Carbon dioxide spewing from the tailpipe of a new car rolling off the assembly line in Detroit could one day alter temperatures in Melbourne, Paris, and Beijing. The face of the tropics is changing on a global scale. In 1980, the United Nations Food and Agriculture Organization estimated that 28 million acres of tropical forests were being lost each year. Ten years later, satellite images suggested that the estimate was low. The World Resources Institute estimated in 1990, using all available data, that the loss was closer to 40 million or 50 million acres per year.[2]

The question is: Are these changes threatening the global ecosystem? Or is the system resilient enough to cope with these changes, adapt, and continue supplying us with air, water, and food?

"All we have to do is check the skies every spring to see the drastic decline of songbirds," Eldredge wrote. "Migrating songbirds fill the same role on a global scale that those caged canaries used to perform for miners. By now, it is abundantly clear to all who will look that the global miner's canary is not at all well."

The fossil record provides evidence of six catastrophic extinctions during the past 4 billion years. The first, and the least severe, came at the end of the Cambrian period, about 510 million years ago. Half the families of trilobites found before that time disappear from the fossil record afterward. Each family contained many species. The disappearance of all those species is evidence of some dramatic occurrence on earth, but what?

Paleontologists cannot tell for certain. One problem is that it is difficult to determine how long it took for the trilobites to become extinct. What looks like a moment in geological time could have taken years, thousands of years, or even several million years. In some areas, fossil trilobites seem to disappear gradually through a series of strata. One suggestion is that a rise in sea level altered the shoreline habitat enough to wipe out many of the species. It could have been the result of some climate change. But the answer will probably never be known.

More severe mass extinctions occurred again at the end of the Ordovician period, 439 million years ago, and at the end of the Devonian, 362 million years ago. The biggest of them all occurred at the end of the Permian, marking the change from the Paleozoic era to the Mesozoic. As many as 96 percent of all the species then living on earth became extinct. The three later extinction episodes—including the most recent, which exterminated the dinosaurs 65 million years ago—pale in comparison.

Again, however, paleontologists are unable to say for certain what happened. As with the extinction that ended the Cambrian period, the devastation could have come all at once or in a series of gradual extinctions over a short period of geological time. Again, however, some speculation leads to the idea of a change in climate.

Eldredge, in his review of mass extinctions, concludes that evidence is growing that global climate change, most likely global cooling, could have been a critical factor. A key insight, however, is that climate might operate indirectly by altering habitat rather than killing species directly. Individual animals vary in their sensitivity to climate, but when habitat is destroyed, it can take with it a broad range of species, possibly producing a mass extinction.

For most of their history, humans and their ancestors were buffeted by ice ages and other climate changes as they gradually spread across the globe. Unlike many other species, humans were generalists who could live in a wide variety of habitats. About eight thousand or nine thousand years ago, however, something changed. Humans stopped responding to change and became agents of change.

At that time, North America was inhabited by giant bisons, elephants, and rhinoceroses. It was about the time that farming was appearing throughout the world. But humans were still hunters, as they had been for thousands of years. With their primitive weapons, the hunters had developed a variety of tricks to help them kill beasts that outweighed them and could easily outrun them. Herds were driven into blind corners or forced over cliffs.

Some researchers believe that the hunters were responsible for the disappearance of several large North American mammals. Not all agree. Climate could have been responsible, too, and other large mammals had disappeared early, with no help from humans. But there is at least the possibility that humans had by then begun to alter their environment in ways they never had before.

An even more important change was then just beginning, however. Hunters eliminated several species, not all. They were selective. But the next change was less discriminating. That was the change to agriculture.

Paleontologists speculate that habitat loss may have been a factor in prehistoric mass extinctions. At roughly the same time that humans may have been driving some North American mammals to extinction, they were also beginning to alter their habitat for farming. "With large-scale habitat alteration largely for agricultural purposes, *Homo sapiens* began to mimic the effects" of the prehistoric mass extinctions, Eldredge concluded. "For the first time, we have a species on earth that has been altering habitat so pervasively that the effects even now border on true mass extinction."

Mass extinctions served a purpose throughout evolutionary time. Each extinction left countless ecological niches unfilled. That provided the opportunity for new species to evolve to fill them. If the end of the twentieth century marks the beginning of another mass extinction, this one caused by humans, that would open up possibilities for new kinds of life. In the long run, is that not how evolution works?

After each of those extinction episodes, the replenishment of life took millions of years—25 million years after the Ordovician extinction and 30 million after the Devonian. After 96 percent of species perished in the great Permian extinction, it took 100 million years for life to recover. "That should give pause to anyone who believes that what *Homo sapiens* destroys, nature will redeem," Wilson said. "Maybe so, but not within any length of time that has meaning for contemporary humanity."

Eldredge recalled an anecdote concerning the great anthropologist Margaret Mead. On her deathbed, Mead looked up and said, "Nurse, I think I'm going."

"There, there, dear, we all have to die sometime," the nurse said.

"Yes, but this is different!" Mead snapped.

So, too, with the ecosystem we live in, Eldredge said. "This is not just any ecosystem—this is our ecosystem, our very own existence as a species, that is at stake," he said. Yes, intriguing new forms of life might repopulate the earth after humans have disappeared. "But that's later," Eldredge wrote. *"This is now."*

A GLOBAL CENSUS

Wilson has done as much as anyone to try to assess the status of the current rate of extinction. In the 1960s, when biologists were asked how many species lived on earth, they routinely answered, "One million." That was the number that had been cataloged. Some suspected that another million or two remained to be found. That was guesswork. As Wilson pointed out, biologists have scarcely begun surveying the planet. When Wilson set out to determine the rate of extinction, his first job was to make a better estimate of the number of species on earth now.

Wilson has spent much of his life roaming the tropics. There life is garishly displayed in all its richness and complexity. The teeming diversity in the tropics is fueled by a scant 10 percent of the sun's energy. That is the amount captured by plants in photosynthesis. That energy is transferred up the food chain to insects and herbivores, from there to spiders and other low-level carnivores, and eventually to birds, mammals, and sharks—the top carnivores, as Wilson calls them.

Humans, too, are at the top of that chain. Like every other living thing on earth, humans depend on solar energy. But animals cannot harness that energy without green plants. Plants are solar-energy collectors. Their role is to convert solar energy into a form that herbivores can use. Humans derive their energy calories directly by eating plants or indirectly by eating animals that have eaten plants.

Biologists have a reasonably good understanding of the origin of biological diversity. Evolution and natural selection have combined to produce an astounding diversity of life. Studies of the origin of species do not allow predictions of the numbers of species, however. "In the realm of physical measurement, evolutionary biology is far behind the rest of the natural sciences," Wilson said. Physicists know the mass of an electron with great precision. They know the diameter of the earth, the speed of light, even the approximate number of stars in the Milky Way. Biologists, on the other hand, cannot say whether the number of species on earth is closer to 10 million or 100 million.

Scientists have described about 1.4 million species, according to Wilson's estimate. That number could be off by 100,000, he said. The largest single group is insects, with 750,000 species recorded. They represent more than half of all known species. Forty-one thousand vertebrates and 250,000 plants have been described. At least 75,000 of the plants are edible. The remaining species are invertebrates, fungi, algae, and microorganisms. Three thousand species of bacteria, 30,800 species of protozoa, 28,600 species of cup fungi, and 500 species of plasmodial slime molds have been described.

Many new species are discovered every year, however. And of those already discovered, 99 percent are known only by a scientific name and a handful of specimens stored, like the imperial woodpecker, in museums. "It is a myth that scientists break out champagne when a new species is discovered," Wilson said. "Our museums are glutted with new species. We don't have time to describe more than a small fraction of those pouring in each year." An enterprising biologist can take a plane to the tropics and find a new species in a matter of days, sometimes within a few hours.

The question, then, is: How many more of these undiscovered species are out there? To make an estimate, Wilson began by thinking about insects, by far the most numerous of the known species. Insects and other arthropods—which include spiders, crustaceans, and centipedes—make up about 875,000 species. They are so important to the world's ecosystems that if they disappeared, humans probably would not last for more than a few months, Wilson said.

As long ago as 1952, an Agriculture Department scientist estimated the actual number of insects at 10 million. In 1982, Terry Erwin of the Smithsonian Institution went to the Peruvian Amazon to collect insects. He devised a method for "bombing" a tree with an insecticide and collecting the insects that fell to the ground. That enabled him to assess the numbers of insects in the rain-forest canopy, a largely unexplored realm.

Erwin estimated, for example, that there were 163 species of beetles living in a single species of tree. Beetles represent only 40 percent of all insects. And the tree he studied was one of some fifty thousand species of tropical trees.

If similar numbers of beetles and corresponding numbers of other insects live in other tree species, then there must be about 20 million species of arthropods *in the rain-forest canopy alone,* Erwin estimated. As a rule, there are twice as many species in the canopy as there are on the ground, so the total number of insect species in the rain forest could be as high as 30 million. (Wilson once found forty-three species of ants on a single tree in Peru, about the same number known in all of the British Isles.)

More recent studies have suggested that Erwin's estimate might be too high. The tree Erwin first looked at might have sheltered an unusually large number of species, skewing and inflating the total estimate. Those other studies suggest that there may be 5–10 million insect species in the tropics, which still dwarfs the 750,000 species known.

Subsequent studies have shown higher than expected numbers of lichens, fungi, roundworms, mites, protozoans, and bacteria in the upper reaches of the rain forest, suggesting that the number of those organisms has also been underestimated.

Not all of the arthropod diversity is located in the tropics. In 1991, Andrew Moldenke of Oregon State University estimated that the H. J. Andrews Experimental Forest in the old-growth forests of Oregon is home to about eight thousand species, most of them insects and other arthropods. The official tally of identified species at the station, run by the U.S. Forest Service, is 143. Most of the rest are unknown to science. "We've come to suspect that these invertebrates of the forest soil are probably the most critical factor in determining the long-term productivity of the forest," Moldenke said.[3]

Some habitats—such as the seafloor and tropical soils—have not been explored enough to allow even rough estimates of the number of species to be found there, Wilson said. Scrapings of the deep seafloor have revealed far more species in that cold, dark place than biologists had expected. There are likely to be at least hundreds of thousands of species there. When bacteria are included, the total could be in the tens of millions.

In February 1994, Marjorie Reaka-Kudla released a study estimating that coral reefs were home to about 425,000 strange and exotic plant and animal species, 90 percent of them unknown to science. Coastal reefs cover roughly 230,000 square miles around the world, and they have not been widely studied, Reaka-Kudla said.

"I claim there is high undocumented diversity," she said. "Some people are saying the oceans are not as vulnerable to extinction" as the land. But she believes "we have underestimated the potential for extinction" partly by underestimating the number of species. "The risk of extinction is extremely high in marine environments." It may be particularly devastating in the oceans, in fact, because they contain more isolated groups of species that have evolved along divergent paths for hundreds of millions of years, she said. "The possibility in marine organisms is for completely unsuspected things because of this ancient path of evolution they've undergone."

She derived her estimate of the number of reef species by assuming that the biological relationships found in rain forests are matched by similar processes in coral reefs. She then took account of the total extent of the world's reefs, compared it to the total extent of rain forests, and calculated the number of species as a fraction of the number found in rain forests.

The number of reef species known to scientists is somewhere between thirty-five thousand and sixty thousand, Reaka-Kudla said. Most of the unknown species are likely to be algae, corals, fishes, and tucked inside the reefs' nooks, worms, clams, snails, sea urchins, and crustaceans. "These are the equivalent of the small rain-forest animals that are so diverse," she said.

Unfortunately, however, corals are declining drastically around the world, Reaka-Kudla said. The principal causes of reef destruction are sewage discharge and soil erosion, which flood the reefs with nutrients. The nutrients in turn promote the growth of algae, which overwhelm and destroy the fragile reef ecosystems.

The thinning of the atmosphere's ozone layer, which absorbs much of the sun's ultraviolet radiation, may also play a role in coral-reef destruction, she said. Corals are sensitive to ultraviolet radiation, which is reaching the earth's surface in higher amounts as the ozone layer undergoes damage from chlorofluorocarbons and other man-made chemicals.

Rita Colwell of the University of Maryland has found that the open ocean is teeming with previously unknown viruses. The viruses are so numerous that they make up a significant portion of the total biomass, or living stuff, in the ocean, Colwell said. They are so different from anything else known that researchers have not yet figured out how to grow them in the laboratory. "We're now just beginning to get a glimpse into this wonderful diversity that is functioning, that we must not destroy," Colwell said.

The study of bacteria and viruses is critically important to medicine, but even so, biologists have no idea how many species of microorganisms might exist on earth. "Take a gram of ordinary soil, a pinch held between two fingers, and place it in the palm of your hand," Wilson said. "You are holding a clump of quartz grains laced with decaying organic matter and free nutrients and about ten billion bacteria."

But how many species are in that gram of soil? Wilson points to a Norwegian study in which researchers looked for unmatched strands of DNA in a gram of soil from a beech forest. Each unmatched strand would indicate a distinct species. In that single gram of soil they found four thousand to five thousand species. In a soil sample taken from sediment just off the Norwegian coast, they found another five thousand species—ten thousand altogether in two grams of soil.

In 1986, Wilson ventured an estimate that there were between 5 million and 30 million species on earth. In his more recent writings, he has abandoned estimates. When microbes are included in the census, the total is probably closer to 100 million than to 10 million. There is no way to know.

The problem becomes even more complex when biologists try to assess biological diversity within a species. That is what interests plant breeders. Corn is not going to become extinct, but thousands of varieties of corn and its wild relatives are at risk. The genes needed to combat the next corn epidemic might be found in only a handful of those thousands of varieties.

Even within a single variety there are genetic differences between one individual plant or seed and the next. The same is true for human beings: No one individual carries the genetic diversity that is evident in human beings. A roomful of people capture more of the diversity, and a stadium crowd captures even more. But even a Superbowl crowd does not capture more than a sliver of the human genetic diversity found on other continents.

Preserving representative populations of species is an even more daunting task than trying to preserve the species themselves. Estimates of the numbers of species on earth are dwarfed by estimates of populations. Wilson noted, for example, that about ten thousand species of ants have been identified. But an estimated 1,000 trillion ants are alive at any one time. That is about two hundred thousand ants per each human being, or 1 million ants for a family of five on a picnic. The wealth of genetic diversity within the ant population alone is staggering.

Another indication of diversity within species is the number of genes a given species has. Flowering plants have as many as four hundred thousand genes. Each can vary from one individual to the next. The number of possible combinations of those genes is almost limitless. Each individual flowering plant carries an array of traits that may be unique, although each of the individual traits would be shared by many other plants. When the number of species is multiplied by the number of genetic traits in each species, the numbers become astronomical.

DOCUMENTING THE LOSSES

If Wilson found it difficult to estimate the number of species remaining on earth, he had even less success trying to estimate their rate of disappearance. "I cannot imagine a scientific problem of greater immediate importance for humanity," Wilson said. He called extinction "the most obscure and local" of all biological processes. In order to know that a species is extinct, researchers have to know where it lives and where it hides. Since so many are undiscovered, it is impossible to know how many are going extinct.

Wilson did not stop there, however. It is possible to say something about extinction rates among species that are known to science. Among that small fraction of the world's species, "extinction is proceeding at a rapid rate, far above prehuman levels," Wilson found.

He listed some of the evidence. About 2,000 species of birds have become extinct in the past two thousand years, dropping the total now to 9,040. Eleven percent of those are endangered. The population densities of migratory songbirds—Eldredge's global "miners' canaries"—have declined 50 percent since the 1940s.

Twenty percent of the world's freshwater fish species are extinct or threatened.

Donald Falk, director of the Center for Plant Conservation at the Missouri Botanical Garden in St. Louis, has estimated that three thousand to five thousand of the twenty-five thousand plants native to the United States are in danger of extinction. Only 250 of those are listed as endangered or threatened by the U.S. Fish and Wildlife Service. But of those 250, at least 37 are potential sources of germplasm for agriculture, said Calvin Sperling, a plant explorer with the U.S. Department of Agriculture.[4]

Speaking before a congressional committee in May 1989, Falk said: "Among the endangered species of North America, there are close endangered relatives of major grains, vegetables, fruits, oil seed crops . . . endangered plums, raspberries, blueberries, squash, corn, tobacco, sunflowers, wild rice, as well as familial relatives of a wide range of other crops.

"Among the native plants of the U.S. there are crop relatives adapted to alkaline or acid conditions, poor soils, and drought: precisely the conditions with which American agriculture may have to contend in coming decades if current predictions of global warming are accurate."

Falk asked the congressmen to think of an ecosystem as a building. "Each species is like a brick in the foundation. They are the building blocks of more complex systems, and it is their particularly adapted sizes and shapes that make the structure so sound. Every time a species becomes extinct, then, is like pulling a brick out of the foundation. You can probably survive a few such losses. But by the time twelve percent to twenty percent of the bricks have been pulled out, the whole structure is going to become dangerously unstable."

In 1992, the United Nations Food and Agriculture Organization (FAO) estimated that forty thousand plant species would become extinct by the middle of the twenty-first century. "Their loss constitutes a grave threat to our world food security," said Edouard Saouma, the FAO's director general.

Wilson noted the importance of the disappearance of entire

ecosystems, the wilderness equivalent of the "genetic wipeout" that Wilkes worries about with crop germplasm. Madagascar is a classic example. Because of its long isolation, almost all of its plants and animals are found nowhere else. That is true of 80 percent of its ten thousand plant species, which include a thousand kinds of orchids. Slash-and-burn agriculture by Madagascar's growing population is rapidly destroying what remains of the forest. When it goes, thousands of species will go with it.

The Atlantic coastal forest of Brazil is another example. As Russell Mittermeier of Conservation International has documented, less than 5 percent of the original forest remains. It is the home of the majestic muriqui monkey, the largest monkey in the New World; the golden lion tamarin; and many other plant and animal species that are found nowhere else. As in Madagascar, population growth and agriculture were mostly responsible for the destruction of habitat.

Wilson found a few studies that suggested that 15 percent of the plant species in South and Central America are likely to disappear within the next century. That will include many of the important wild relatives of beans, potatoes, peppers, tomatoes, corn squash, peanut, cacao, and peppers, all of which originated there. Previous mass extinctions mostly involved animals. This is the first time plant diversity has declined sharply, he said.

Wilson could not tally current species losses directly because so little is known about the number of species on earth now. So he fell back on a principle he helped devise early in his career. Wilson studied populations of species on islands and noted what happened when a habitat on those islands was reduced. He found, as a rule of thumb, that when a habitat is reduced to one-tenth its original size, half of the species that were there become extinct.

Tropical rain forests now cover an area about the size of the continental United States. That is about half the area they covered in prehistoric times. Half of the primeval rain forests are already gone.

They continue to decline at a rate of about 1.8 percent per year. Using Wilson's formula, that represents a decline of about one-half of 1 percent of the rain-forest species each year.

"If destruction of the rain forest continues at the present rate to the year 2022, half of the remaining rain forest will be gone. The total extinction of species this will cause will be somewhere between 10 percent . . . and 22 percent," Wilson calculated. That translates to 5–10 percent of the world's species, possibly more.

Continuing the arithmetic and using a conservative estimate of 10 million species in the rain forest, Wilson determined that twenty-seven thousand species become extinct each year. "Each day it is seventy-four, and each hour it is three," he said.

The expected rate of extinction, without human intervention, would be about one species per million per year. "Human activity has increased this extinction between one thousand and ten thousand times over this level in the rain forest by reduction in area alone.

"Clearly," Wilson ruefully concluded, "we are in the midst of one of the great extinction spasms of geological history."

Wilson is one of the few ecologists to take notice of the potential effect of this loss on agriculture. In addition to the loss of wild relatives of established crops, the disappearance of natural areas in the tropics is leading to the loss of countless other plants that could become staples.

Camu camu, one example from the Peruvian Amazon, has up to thirty times the vitamin C of citrus fruit. The winged bean, a native of Southeast Asia, is known as the "supermarket on a stalk," according to Noel Vietmeyer of the National Academy of Sciences, an expert on underutilized tropical plants. Everything on the winged bean is edible. "The leaves are like spinach, the pods are like green beans, the tendrils are like lacy asparagus," he said. The roots are nutty-tasting tubers with four times the protein of potatoes. The seeds are as much as 42 percent protein, and the vitamin A content is among the highest ever recorded.[5]

There are many other examples, some of which are now entering the American market. Some supermarkets now carry as many as four hundred different kinds of produce during the course of a year, Vietmeyer said. Many have arrived in the United States to satisfy the tastes of immigrants from the Caribbean, South America, and Southeast Asia.

The loss of genetic diversity in the tropics, while it can take with it wild relatives of corn, wheat, and other major crops, is also threatening many other crops that are essential foods in other countries. In 1990, Robert and Christine Prescott-Allen, the two conservationists from British Columbia who assessed the economic value of wild germplasm, did a study that found that humans rely on a far broader variety of crops than is usually recognized.

Many researchers had concluded that humans survive on an extremely narrow range of crops. In 1987, the Office of Technology Assessment (OTA) noted that 75 percent of human nutrition is provided by only seven plants: wheat, rice, corn, potato, barley, sweet potato, and cassava. That is true in terms of global totals. But it overlooks such foods as olive oil, the leading source of plant fat in Greece and Italy, or sunflower oil, the leading source of plant fat in several European and South American countries. The Prescott-Allens concluded that human diets rely heavily on 103 species of plants, not the 7–30 species suggested in previous studies.

The problem with focusing on the short list of foods that supply most of the world's calories is that it makes all the other foods seem of minor importance. The implication, the Prescott-Allens said, is that the world can survive on only a few crops and that protecting the genetic diversity of those crops is enough to protect the world's food. In fact, the job is much larger than that, they said.

"The lesson is that we should be making a much greater effort to conserve a broader range of economically useful species, and their relatives, even though they may not be part of the top thirty," Robert Prescott-Allen said.

Wilson, Hugh Iltis, and many others have urged that exploration and preservation of the world's biological diversity be accorded top scientific priority. "It is . . . imperative that we study and carefully preserve nature on this planet now, for this will be our last chance to ensure that biodiversity will survive for future generations," said Iltis.[6]

A wide variety of conservation groups, biologists, and government agencies have undertaken limited biological surveys in the tropics. But researchers have not had the resources for large-scale, worldwide surveys. At the UN Earth Summit in Rio de Janeiro in

1992, the nations of the world overwhelmingly approved an international agreement to protect biological diversity. It is not clear, however, what the practical effect of the agreement will be.

While seed banks can be helpful in protecting certain crop varieties important for agriculture, they can only do a small part of the job, Wilson said. "All the efforts of the existing seed banks to date have been barely enough to cover a hundred species, and even those are in many cases poorly recorded and of uncertain survival ability," he wrote in *The Diversity of Life*. "*Ex situ* methods will save a few species otherwise beyond hope, but the light and the way for the world's biodiversity is the preservation of the natural system."

THREATS TO SEED BANKS

A further difficulty with seed banks is that they are vulnerable to destruction during political upheavals or natural disasters. Twice in the 1970s, Nicaragua lost many of its valuable corn samples: once after an earthquake and again during the Sandanista revolution. Traditional varieties stored in the Peruvian village of Huancapi were destroyed in a guerrilla attack. Workers at a potato gene bank in Peru, protesting low wages, ate the national collection of potato varieties. Seeds of rice landraces were eaten or rotted in Cambodia during the political strife of the 1970s.

The Nicaraguan collection was restored with the help of Mexican researchers who had kept duplicates of many of the Nicaraguan samples in the collection at the International Center for the Improvement of Maize and Wheat (CIMMYT) in Mexico. Duplicates of the Peruvian potatoes were stored in a seed bank in Bolivia. But in many cases samples cannot be restored. Some of the Cambodian landraces were returned to the country from the collection at the International Rice Research Institute in the Philippines; others were lost.[7]

Another near miss occurred early in 1991 in Ethiopia, when rebel guerrillas invaded the capital, Addis Ababa, where the government maintained a seed bank with an important collection of wheat, barley, and sorghum germplasm.

"Ethiopia is a secondary source for wheat and one of the

prime centers of origin for sorghum," said Steve Eberhart of the National Seed Storage Laboratory (NSSL). "We've gotten some very useful sorghum from them over the years. The last time they had the war [when Communists threw out the monarchy], it pretty well wiped out the gene bank there, and we sent back a lot of materials to them."

A special problem with Ethiopia was that it restricts export of its germplasm; so many of the samples in the seed bank were not duplicated anywhere else, Eberhart said. As fighting neared the capital, Melaku Worede, the director of the gene bank, quickly contacted U.S. officials to see whether it might be possible to move the collection to the United States for safekeeping.

Despite Ethiopia's ban on exports, Worede thought that public pressure might persuade the government to let the seeds out of the country. Before the plan could be put into effect, the rebels captured Addis Ababa, and the fighting ended. The seed bank and its collection were not damaged.

More recently, the bloody conflict that disrupted farming in Somalia led to widespread starvation. Somalia's two seed banks were looted, and the seeds were eaten. Hundreds of varieties of crops adapted to Somalia were destroyed. Fortunately, about three hundred varieties from the Somali collection had been duplicated in Kenya's seed bank. The International Board for Plant Genetic Resources (IBPGR) in Rome immediately made plans to produce more seeds for return to Somalia.[8]

International upheaval or natural disasters are not the only causes of genetic wipeout, however. Many distinguished plant breeders have built up extensive seed collections during the course of their careers, only to have the seeds literally thrown away after their retirement. "The U.S. has no policy, no clearinghouse, for privately and/or publicly held research and working collections of genetic stocks," said Wilkes.[9]

THE NORTH VERSUS THE SOUTH

The huge dependence of the United States and Europe on germplasm from the tropics has led to a smoldering diplomatic dispute between developed countries and developing countries over germplasm ownership. The industrialized countries are in a better position than the poorer countries to exploit germplasm. The developing countries are increasingly angered by the exploitation, by someone else, of their resources—for which they receive no compensation.

Some countries have gone so far as to ban the export of seeds. That is a difficult policy to enforce, as Italian rice farmers found out when Thomas Jefferson slipped out of Italy with a pocketful of their seeds. But prohibitions on germplasm export could disrupt what has been, until now, a fairly open system of exchange among countries.

Export bans could also threaten the development of in situ reserves that are supposed to serve as global repositories for crop germplasm. Without a free flow of germplasm across national borders, countries outside the centers of crop diversity will have no alternative to preserving germplasm in seed banks. The advantages of in situ reserves will be lost to them.

The dispute that threatens germplasm exchange is usually referred to as the north-south debate. Most of the world's crops—and crop germplasm—originated in what is loosely referred to as the south: the developing countries of the tropics. Most of the exploitation of those resources, through the development of modern crop varieties, occurs in the industrialized countries of the north.

This unequal distribution—of resources in the south and technology in the north—suggests the opportunity for collaborative enterprises that could benefit people on both sides of the equator. But that is not the way it happened. While the countries of the south supply the raw materials that fuel the commercial seed enterprise, the countries of the north collect the profits.

Many breeders argue that crop germplasm should not be owned by anyone. Unlike other natural resources, such as oil or minerals, germplasm ought to be treated as the "common heritage of mankind," many breeders say. It belongs equally to the nations of the south, where it originated, and to the breeders of the north.

Yet activists and some officials of developing countries see that viewpoint as one that tips all the advantages to the north. In the simplest scenario, the north is taking the south's germplasm, using it to produce high-yielding crop varieties, and selling the seeds of those varieties back to the south.

In practice, the situation is more complicated than that. Putting genes from wild crop relatives and landraces into modern crop varieties requires two or three decades of breeding. Seed companies do not usually work directly with wild or traditional material; they work with more advanced breeding lines. But those breeding lines often contain genetic traits that originated in plants harvested from developing countries.

Commercial varieties incorporating those traits may be highly profitable for seed companies. But developing countries, despite their critical contributions, get nothing. Indeed, if they buy the seeds, they are, in effect, paying for germplasm that was theirs to begin with.

That is where the perceived inequity lies: Germplasm is free when it is moving from developing countries to the seed companies of the United States and Europe. But when the germplasm returns to the south, it has a price tag on it.

Seed companies clearly need to be compensated for their research and development efforts. The germplasm that may be sold back to the south is a vastly improved and refined version of what was harvested in the south. But that sidesteps the issue of whether the south ought to be compensated for its contribution as well.

The problems of restrictions on germplasm exchange has now extended to publicly financed breeding programs. According to the crop scientists Major Goodman and Fernándo Castillo-Gonzalez, breeding lines developed by government programs in Mexico, Peru, and Yugoslavia are not readily available to others. "Recently, even a few publicly supported U.S. breeding efforts, such as the corn-breeding program at Cornell University, have restricted access to their breeding materials," the researchers wrote. "Other public programs are considering sale of exclusive rights or royalty payments for access to lines they have developed."[10]

Jack R. Kloppenburg, Jr., a University of Wisconsin historian, has written extensively on the north-south debate. He and others have argued that developing nations ought to get some kind of compensation for their contribution to modern crop varieties.

"There is a great irony in the germplasm controversy," he wrote with a colleague several years ago. "In a world economic system based on private property, each side in the debate wants to define the other side's possessions as common heritage. The advanced industrial nations of the North wish to retain free access to the developing world's storehouse of genetic diversity, while the South would like to have the proprietary varieties of the North's seed industry declared a similarly public good."[11]

As Kloppenburg noted, seed companies have several objections to paying developing countries for their germplasm. One is that germplasm is of indeterminate value until it is evaluated. Setting a price on wild-crop relatives and landraces would be difficult. Some breeders would argue that germplasm does not have any substantial value until it has been improved and incorporated into advanced, modern varieties.

A second objection is that taking germplasm from a country does not deprive the country of anything. The harvest of a few handfuls of seed does not prevent the country from harvesting seed and devising its own improved varieties.

Kloppenburg admitted that it would be difficult to allow markets to put a price on germplasm. But it should be possible to negotiate agreements of some sort to set prices, he said. In response to the argument that germplasm is worth little until it is improved, he observed that peasant farmers have been improving germplasm for centuries and thus have as much right to compensation as do scientific breeders in seed-company laboratories. Lastly, Kloppenburg argued that developing nations do indeed lose something when they share their germplasm. They lose the exclusive right to exploit the germplasm, meaning that they are likely to lose potential markets to competing nations.

Ethiopia is one country that has reacted angrily to what it perceives as inequities in the international exchange of germplasm. It now strictly controls export of its important collections of wheat, coffee, and other crops. That policy almost backfired when the Ethiopian collection was threatened by armed rebellion. If the collection had been destroyed, there would have been no way to replace many of its samples, because they are not held in reserve anywhere else. Nevertheless, Ethiopia has insisted that its valuable crop genetic resources be kept within its own borders.

Much of the debate over the international exchange of germplasm has taken place at meetings of the United Nations Food and Agriculture Organization. One of the most prominent voices there has been that of Pat Roy Mooney, a Canadian economist who has largely shaped the debate with a series of books and articles, beginning with "The Law of the Seed," which he published in 1983.

"For some years now, a kind of gene drain has been under way, siphoning off the Third World's germplasm to 'gene banks' and breeding programs in the North," Mooney wrote. "Germplasm is the raw material of seeds—and seeds are the first link in the food chain. Some governments and some chemical companies recognize this, and a grab is being made for the control of germplasm."

Agricultural development in poor countries will be severely hampered, Mooney argued, "if our seeds are subject to exclusive monopoly patents and our plants are bred as part of a high-input chemical package in genetically uniform and vulnerable crops."[12]

Mooney and others see the international germplasm conservation effort as a veiled movement by industrialized nations to gain control over the south's germplasm resources. Much of the international effort is coordinated through the IBPGR in Rome. The board is part of the Consultative Group for International Agricultural Research (CGIAR) at the World Bank. That is the international network of agricultural research facilities that also includes the corn and wheat research center CIMMYT in Mexico and the International Rice Research Institute in the Philippines.

The plant genetic resources board advises developing countries on the collection and preservation of germplasm. Some nations have established seed banks in conjunction with CGIAR's international agricultural research centers. Others have established national seed banks. But the IBPGR's contribution has been far too small to assure that the seed banks meet minimum standards, Goodman and Gonzalez have said.

Using language similar to the language they used to describe American seed banks, the researchers wrote, "IBPGR's budget . . .

precludes continuous support of national or regional programs. The result is that many, if not most, IBPGR financed germplasm repositories are more accurately described as potential seed morgues rather than seed banks."

CIMMYT, one of the major contributors to the Green Revolution and one of the best run of the international agricultural research centers, is dreadfully behind in the upkeep of its seed collection, Goodman and Gonzalez found. CIMMYT "still regenerates only about 200 accessions per year and has a backlog of at least 2,500 materials that have never been regenerated," they said.

Others have found deficiencies in other seed banks overseas. Laura Merrick of the University of Maine examined the collection of squashes and pumpkins in Costa Rica, one of the most environmentally enlightened Latin American nations. Of twenty-five hundred samples in the collection, one thousand were given the wrong species identifications. And when they were correctly identified, many turned out to be duplicates of other samples in the bank, she said. "After I looked at their collection, I figured there were perhaps only fifteen hundred unique accessions."

Goodman and Gonzalez argued that the IBPGR and the international agricultural research centers are failing to meet the challenge of making better seeds available to farmers in the developing world. "If you travel 10 miles from CIMMYT's impressive headquarters near Mexico City, many local farmers grow the same ancestral varieties of corn that their great-grandfathers grew. Increased agricultural productivity in developing nations depends not on new hybrids, but on access to quality seeds, quality information, adequate water, ready financing, decent markets and fertile soil," they wrote.

"In contrast to social activists' charge that the north is stealing germplasm from the south, the crime is the manner in which both the north and the south are mishandling the world's germplasm resources," the researchers wrote.[13]

In 1983, the Food and Agriculture Organization established what it called an "international undertaking" providing for the free exchange of germplasm. The catch was, however, that commercial varieties were included in the undertaking. A broad range of developing countries signed the document, but developed countries stayed away, knowing that to sign it would mean forgoing breeders' right to profit from their commercial varieties.

Developed countries not only refused to sign the pact, they moved in the opposite direction, toward greater protection of breeders' rights. In April 1988, the U.S. Supreme Court granted patent protection to a genetically engineered mouse produced at Harvard University. According to John Barton, a patent authority at Stanford Law School, that case removed the last obstacle to the patenting of living things.

"The patenting of life-forms is rapidly expanding in the U.S., which now provides intellectual property protection to microbes, plants and animals," Barton wrote in March 1991. In competitive economies, such as those of the developed countries, patents can foster innovation. In developing countries, however, patent protection can lead to monopolies, he said. As a result, many developing countries have excluded food and drugs from patent protection.

In the area of plant breeding, however, the issues become considerably more confused. U.S. law allows the patenting of genes created through genetic engineering. But what about genes found in nature? "Imagine the political outcry if a company discovered a useful disease resistance gene in a natural Mexican weed—and then sought to patent its use in commercial varieties of maize that would be sold back to Mexico," Barton said.

On the other hand, he noted, companies might reasonably argue that they deserve some protection for the effort expended to breed such natural genes into useful crops. The first company to adapt a natural gene to a new use faces large costs. It also faces the threat that once it has demonstrated the gene's utility, others can imitate it quickly and cheaply. "This situation, in which the innovation costs are heavy and the imitation costs are slight, is precisely the one in which patent protection can be most beneficial as an incentive," said Barton.[14]

Proponents of the Food and Agriculture Organization's undertaking on plant genetic resources have come to recognize that some acknowledgment of breeders' rights is necessary. More recently, a series of meetings under the auspices of the Keystone Center have led to a proposal for the establishment of an international fund to aid farmers and breeders in developing countries.

In 1991, Merck & Company, the pharmaceutical giant, signed a $1 million contract with Costa Rica's National Biodiversity Institute giving the company rights to screen tropical plants as potential sources of drugs. The agreement provides that profits Merck earns from plant-derived drugs will be shared with Costa Rica. The agreement could serve as a model for multinational seed companies.

Many observers now agree that the north-south debate must be resolved by some kind of global partnership that recognizes the differing roles and responsibilities of the industrialized nations and the developing countries. "Free access," as one researcher pointed out, is not equivalent to "free of charge."[15]

Chapter 6

The Last Harvest

It is a hot, brilliant sunlit morning in mid-June. The temperature is expected to reach into the nineties later in the day. The spring has been unusually hot and dry, enough to worry Glenn Fritz as he pulls his truck into the backyard of his son's house. In the distance, a few blocky farm buildings dot the horizon. Spidery high-tension towers march away in a solitary column until they vanish. Every other square inch of land under the cloud-dappled blue sky is colored with the vibrant green of young corn plants in seemingly endless parallel rows.

Despite the punishing heat, which has baked the earth and the seedlings for a week, Fritz's crop is healthy. The broad leaves of the corn, now about fourteen inches high, curl backward, fully exposed to the sun's punishing rays. Three days ago, the plants were doused with three and one-half inches of rain in a single day, saving Fritz from a potentially catastrophic loss. "It's been real dry," he says. "We've had six weeks with a half inch of rain. It's the most unusual thing I've seen in forty years of farming. A week ago, I was under stress. We were almost on the verge of drying up." With the coming of the rain, Fritz says, "we were blessed. As much as we got, we know we're good for three weeks."

Only two miles south of Fritz's farm, the rainstorm that washed his crops dropped a scant tenth of an inch of rain. The corn is beginning to suffer.

"They're all roped up down there," Fritz says as he stands at the edge of his farm and looks over his own crop. He is talking about the way the corn's leaves curl up into long, ropelike tubes when they are short of water. The roping up of the leaves slows evaporation and helps shield the plants from the sun. This is not an accident; the trait was bred into the corn to help protect it from drought. "The older varieties sit there with the leaves open," says Fritz. Nevertheless, the roping cannot protect the corn indefinitely. "They're going to need rain in the next seven days, or some of it's going to start dying."

Like most of his neighbors, Fritz has borrowed money to see him through planting. "Very few of us have the capital to put this in the ground without going to the bank. You're talking $100,000 to $150,000 to get it in the ground. And you don't know whether you're gonna get anything. That's gambling. But it's all we know."

Fritz and his son, Gary, grow corn and soybeans (and a few acres of wheat) on two thousand acres outside Joliet, Illinois, an hour's drive and a world away from the lakefront sprawl of Chicago. Fritz looks and talks the way a visitor to Joliet might expect an Illinois farmer to look and talk. He chooses his words carefully and pronounces them in an accent that is difficult to define except that it clearly has little to do with the choppier rhythms of Chicago. He is of medium height and build, with strong, thick hands and deeply tanned skin toughened and weathered by decades of work outside. He is wearing worn jeans and a light blue chambray work shirt. Friendly blue eyes peer out from beneath the brim of his hat.

"A thousand acres is about right for one man," Fritz says. He and his son own four hundred acres of the two thousand they farm, which is spread over fourteen separate tracts. The rest is sharecropped on land belonging to absentee owners. The owner pays half the cost of the fertilizer and seed. Fritz and his son provide the machinery, the fuel, and the work. At harvesttime, he delivers the grain to the elevator.

Fritz, sixty-three, has been farming since he dropped out of school in the eighth grade. Fritz's father had a disabling heart attack, forcing Fritz to leave school and take over the farm. Glenn is the third generation of Fritzes to farm. Gary is the fourth. "I have five sons," Fritz says. "Years ago, all the sons stayed on the farm. One of mine did." Likewise, Fritz was the only one of his father's sons to stay on the farm. "Around here, there aren't too many fathers and sons."

"It's probably the most wonderful place to raise a family," Fritz says. "The air. You're not smelling that exhaust. The neighbors aren't hanging on your door. You can mow your grass at six in the morning if you want." Fritz has made some sacrifices to keep his family on the farm. "The first twenty-seven years my wife and I were together, I worked every winter in town as a welder. I looked forward to it, but I looked forward to spring," he says. "I sent four of the kids through college."

When Fritz took over the farm, Illinois farmers weren't yet using hybrid corn. That meant they could save the best ears of corn for planting the following year. But hybrid corn swept over Illinois after World War II, as it swept over the nation. "The hybrids were just coming in," he says. He was quick to take advantage of them, but many of his neighbors waited and watched before making the same move. "It took twenty years to sell some farmers," Fritz says. For some of those farmers, it was a fatal mistake. They didn't survive.

Fritz, on the other hand, prospered. The adoption of modern equipment was partly responsible. When Fritz began, he used a planter that sowed two rows of corn at a time. Now he plants twelve rows at once. Pesticides, fertilizers, and modern hybrid corn varieties were the key to that prosperity, Fritz says. Without them, "we would not be in business. Any of the farmers that kept the old ways are in trouble." If Fritz's father had continued to run the farm, he would have been one of those who kept the old ways. "My dad would never buy a bushel of seed corn from somebody else," Fritz says. When his father died at the age of ninety-one on Labor Day, 1993, he had never eaten in a restaurant. Nor had he ever taken a bite of pizza. "He wouldn't even try it," Fritz says.

Life has become easier on the farm, Fritz says. "When I was young, we worked from five A.M. until seven at night." Much of the time went into caring for farm animals. "Every farm used to have its own hogs and chickens. Now you can't sell an egg off the farm without inspections." Many farmers, including Fritz, have given up raising animals. The planting season is busy, and harvesttime is worse, with work sometimes continuing around the clock. Summers are devoted to maintaining machinery and painting, but there are fewer buildings and storage sheds to paint than there used to be. "I didn't have much time. But the younger generation," he says, pointing to his son's house, "he's got a boat." Fritz never learned to swim. On one of his rare visits to Lake Michigan he got out too far, was tossed by a wave, and panicked. "If it wasn't for my wife, I would have been gone," he says. "I never got off the farm, never got to a show."

While Fritz has been quick to adopt new technology in the years since he began, he has inherited his father's conservatism on money matters. That is how he escaped the foreclosures that threw so many farmers off their land a decade ago. "A lot of them thought these good years were never going to end. They borrowed too much. When you're in business for yourself, you're gonna have a good year and a bad year. When that good year comes, you put some away."

Fritz's farm is in Will County, west of Chicago and close enough for subdivisions to begin creeping onto land that has been devoted to farming since it was settled. His great-grandfather, who came to the New World from Germany at the age of eighteen, worked in stone quarries until he was able to buy some land in Illinois from the U.S. government. Fritz's father started with 160 acres, and "then he started buying farms," Fritz says. Fritz is president of the Will County Farm Bureau, a farmers' organization that looks after farmers' interests in the county, the state, and in Washington. In 1993, the county lost eight thousand acres of farmland to subdivisions, roads, and forest preserves. "Everyone wants to get out of Chicago," says Fritz. "They come out here. That eight thousand acres will never come back."

Some come to live, others to speculate in real estate. "I'd say fifty percent of the land here is owned by doctors, dentists, and businessmen," Fritz says. "That was a big push about twenty years ago. They started investing." Much of the land is swapped rather than sold. Farmers like Fritz are reluctant to sell and face the prospect of large capital-gains taxes. So they trade. Developers purchase tracts of farmland farther from Chicago and trade them for parcels closer to the city. The trades are usually done in a ratio of six or eight to one: six acres farther away for each acre in the more desirable location. "I had ten acres I traded six for one," Fritz says. "I couldn't afford to sell it. I traded it. No money changes hands."

Fritz lives in a modern brick ranch-style house he built a few years ago, when he turned his house over to his son, who is now principally responsible for running the farm. (Fritz insisted that he did not want his name on a recently purchased truck. His son had it painted to say: Gary Fritz and Dad.) Fritz's new home is a mile and a half from what is now Gary's home. It is across the street from the house Fritz's father was born in at the turn of the century.

Fritz and his son plant about half of their two thousand acres with corn. On that thousand acres, they generally plant about five different varieties. (Most of the corn produced in Illinois is grown for animal feed or corn oil; only a small portion of the land is devoted to edible sweet corn.) "To this day, we test hybrids against each other," he says. "We run a test plot every year. In the test plot, we've got fifteen to twenty varieties. And the few that do best, we buy for the next year."

Seed companies make samples available for Fritz's test plot, which is only a few steps from his son's house. At this stage, early in the season, there is little obvious difference among the test varieties. "You won't tell the difference until harvesttime," says Fritz. "To look at it, it's all field corn." Some of the seeds were produced by Gary for possible sale to other farmers. "He hopes his do best," Fritz says, smiling.

Fritz likes to show visitors his equipment shed, a steel hangar large enough to house a few small airplanes. "This tractor is a 1961," he says, pointing to the smallest vehicle in the shed. "Look at the shape it's in!" he says proudly. The tractor is clean, and its green paint still shines. "Nothing stays out overnight," Fritz says. Nearby is a $56,000 computer-controlled, radar-guided sprayer, used to dispense insecticides and herbicides.

Fritz points across the shed to the combine, used to harvest the corn and beans. "This is a tool we use maybe twenty-five days a year. The rest of the year it just sits there." Its cost: $110,000. The cab is an air-conditioned glass bubble equipped with a mobile radio. Fritz climbs up and sits down. At ten feet above the ground, he has a commanding view. The enclosed cab is a recent innovation. "I used to come in looking like a raccoon—all the dust," he says. Now that the working conditions have improved, Fritz reserves the job of driving for himself. Fritz and his son plant the soybeans and corn from mid-April until mid-May. The beans are harvested in mid-September; the corn, in mid-October. Beneath the combine's cab, heavy steel rollers separate soybeans and ears of corn from the plants. With corn, the combine also removes kernels from the cobs. The corn or beans are then dumped in a two-hundred-bushel bin in the back, all in one continuous operation. Corn must be dried soon after it is harvested. It is brought in, weighed, and put in a wet holding bin for not more than a day to avoid rotting. Then it is put through a corn dryer, a cylindrical heater with natural-gas burners that heat the corn to three hundred degrees. Fritz can dry as many as five hundred bushels per hour.

Then the corn is put into a cylindrical, corrugated-steel corn bin, about two stories high, with a conical steel top. The advantage over older, wooden storage bins is that this one is sealed. In the old bins, "we'd find fifty or sixty mice that had all been eating and urinating," says Fritz. "Now everything's tight. We have no rat or bird droppings." Nothing gets in or out of the steel bins. That's important when selling corn to Kellogg's, say, for cornflakes. A farmer has to submit an affidavit that the corn is free of insects. "You can't lie to them," says Fritz. He knows of one farmer who did—and got caught. "The city people don't realize the quality of the food compared to what it used to be."

Fritz walks over to a corn bin, climbs a ladder that hugs the outside of the bin, opens a hatch near the top, and looks inside. It is about half full of last year's corn. He sometimes holds corn or beans if he thinks the price might be going up. "I've got most of this sold for futures," he says, climbing down.

In a small building next to the corn bin is a pile of empty plastic herbicide containers, triple washed and waiting to be recycled. "Years ago, you know what we did with these? We threw them in the garbage. We didn't know enough about it."

Fritz applies fertilizer when he plants corn. Herbicides are applied when the seedlings are about two inches tall. "That pretty well controls the weeds for the rest of the year," he says. "Then we knife in the nitrogen," when the corn plants are about four inches tall. A tool is used to poke holes into the ground and inject the nitrogen, a critical nutrient. "It's a gas. It turns to liquid when it gets in the ground."

Fritz is a modern farmer. He encourages traditional values in his children, and he continues some traditions, such as his almost loving care of the tractors and other farm equipment. But he has been eager to adopt innovations that promise to boost the productivity of his crops.

He has an open mind about genetic engineering, which has sparked bitter divisions among farmers, especially dairy farmers, who have been debating the virtues and vices of genetically engineered bovine growth hormone. With regard to the genetic engineering of crops, Fritz says, "Maybe they will get it here in time for another expansion."

Farmers are facing increasing pressure to boost their production while treading more lightly on the environment. The environmental concerns are going to limit farmers' options in ways they have never had to face before. The growing U.S. population will force American farmers to continue to increase their yields. In the meantime, the squandering of water and accelerating degradation of topsoil will force farmers to change many of their practices. Increasing calls for restrictions on pesticides and fertilizers, coupled with demands for energy conservation, will likewise narrow farmers' options.

Furthermore, they face uncertain but potentially dramatic threats from even more exotic problems, such as global warming and the depletion of the stratospheric ozone layer. Solving all of these problems is far beyond the reach of any single farmer, or even of any nation's farmers. Yet these problems will, in coming years, have a direct and immediate impact on every American farmer, as well as their counterparts around the world.

Improved crop varieties are one of the principal means of escape from this shrinking environmental bottleneck. The changes that will be forced upon agriculture in the next century will further add to the importance—and the value—of crop germplasm.

THREATS TO FOOD

The United States is the world's leading food exporter. The value of its agricultural exports stands at about $40 billion per year. Because of the dominant role that the United States plays in world agriculture, gains in American agriculture help to feed people all over the world. Yet American agriculture is already operating on a deficit. In order to accomplish the gains it has made so far, it has consumed land, water, and energy faster than they can be replaced.

When the American economy operates at a deficit, there is at least the theoretical possibility that future taxpayers may be able to pay back the overdraft—as politically difficult as that might be. When land, water, and energy are overspent, there is no such possibility, even theoretically. No technology can create new topsoil for the grain belt. No conceivable machinery can restore clean, fresh water to the America's rivers and aquifers.

When agriculture arose ten thousand years ago, the world had a population of 4 million. The world's population is now growing by that amount every ten days. By the next century, the number of people alive at one time on earth will be one-tenth as many as all the people who have lived during the past ten thousand years.[1]

The world's population has already surpassed 5.5 billion. According to population experts Paul R. Ehrlich and Anne H. Ehrlich, that figure is likely to double before the population begins to stabilize. Since 1968, an estimated 200 million people, most of them children, have died of hunger and diseases related to malnutrition. One billion are hungry. "If the food supply is not what it ought to be now," they observed, "what are the implications for a population twice as large fifty years from now?"[2]

Past gains on the farm were made possible not only by crop improvement but also by increased use of fertilizers and pesticides. But fertilizers and pesticides have now made their contribution. In the view of many experts, chemical additives cannot do too much more. Plants can withstand only so much fertilizer, herbicide, and pesticide. Fertilizers and pesticides are made from petrochemicals, the supply of which is not inexhaustible.

David Pimentel, a professor of ecology and systematics at Cornell University, is one of the few scientists who understand both agriculture and ecology. Pimentel was trained as an entomologist, and his early research was on controlling agricultural insect pests. "I was one of these spray people," he said in an interview in his Cornell office.

He quickly became unhappy with what he saw as the overuse of pesticides, and he has since devoted most of his career to exploring the relationship between agriculture and the environment. In February 1993, Pimentel summed up much of his research in a study that forecast a grim future.

He concluded that land, water, and energy are being used up so quickly that by the year 2100 the population must be slashed to 2 billion or less to provide prosperity for all. The alternative, Pimentel concluded, is a population of 12 billion to 15 billion people and an apocalyptic worldwide scene of "absolute misery, poverty, disease, and starvation."

"There's no way out of it," Pimentel said. "There are just insufficient resources for these people to live like we do today." In the United States, the population would reach 500 million, and the standard of living would decline to something only slightly better than that of present-day China.

Depletion of coal, oil, and natural gas, along with uranium reserves, is one factor limiting the number of people that can prosper on earth. The other two key limiting factors are cropland and water for irrigation, he said. Each of the three factors, considered separately, leads to a calculation that the earth can support only 1–2 billion people at a level comparable to the current American standard of living, Pimentel found.

That is not quite as drastic as it sounds. To reach a population of 2 billion by 2100, families around the world would have to limit themselves to an average of 1.5 children each, a reduction in population growth already achieved in Germany and Italy.

"It doesn't mean no reproduction during this period," Pimentel said. "What needs to be done is for people to make up their minds in a democratic way that we would all like to live with a high standard of living." But it will be difficult in some parts of the world—in many African countries, for example—where population is doubling every twenty-five years. "There's no way out of it," Pimentel said. "There are just insufficient resources for these people to live like we do today."

Paul and Anne Ehrlich took a different approach in estimating the optimal population for the United States. Based on the huge amount of resources each person in the United States consumes, the United States can in some respects be considered the most overpopulated nation on earth, they concluded. America's 250 million people, making up about 5 percent of the world's population, consume 25 percent of the world's fossil fuels. Americans are among the largest users of the chlorofluorocarbons (CFCs) that are thinning the ozone layer. They have destroyed most of their forests.

"Like most of the rest of the world, the United States has been consuming environmental capital—especially its deep, fertile soils, ice-age ground water, and biodiversity—and calling it growth." The optimal population of the United States depends largely on what kind of social system and standard of living is considered desirable, the Ehrlichs said.

The social carrying capacity of the country, which takes those factors into account, is going to be less than the maximum biological carrying capacity. But that maximum figure "implies a factory-farm lifestyle that would be not only universally undesirable but also unattainable," said Paul Ehrlich and Gretchen C. Daily of the

University of California, Berkeley.[3] It is the same thing Pimentel had in mind when he based his calculation of an optimum world population on the current American standard of living. A figure based on the current Chinese standard of living would clearly be higher.

After juggling the numbers and making assumptions about what constitutes a desirable standard of living, they concluded that the optimum population for the United States was probably 75 million.

"That was about the population at the turn of the century, a time when the United States had enough people for big, industrial cities and enough wilderness and open space that people who wanted it could still find real solitude," they said. "With about that number, we believe a permanently sustainable nation with a high quality of life could be designed—if it were embedded in a world that was similarly designed."[4]

Not all of the experts agree with the disturbing projections made by researchers like Pimentel and Ehrlich. Another group of experts believes that improvements in agriculture will allow the food supply to keep up with demand. The two camps divide over the issue of future technology—whether it will or will not provide the increases in food needed to sustain a population expected to reach 10 billion by the year 2050. The pessimists are most often biologists and environmentalists. The optimists are often economists.

The case for the optimists was summarized nicely in the March 1994 issue of *Scientific American* by John Bongaarts, director of research at the Population Council in New York. While the pessimists point to the gloomy population over the next few decades, the optimists attest to the demonstrable growth of the food supply over the past few decades.

Between 1965 and 1990, the calorie intake per person increased 21 percent in the developing countries, from 2,063 calories per day to 2,495, according to Bongaarts. The amount of protein in the diet increased from 52 grams per day in the developing countries to 61 grams daily.

Laboratory research has shown that typical agricultural production is still far below what is possible despite the gains since World War II. One study mentioned by Bongaarts suggests that an additional 5 billion acres of land is available for cultivation in ninety-three developing countries.

Only 34 percent of the seeds planted during the mid-1980s were of modern, high-yielding varieties. More widespread use of those seeds would boost yields. (Bongaarts does not note, however, that this would speed the disappearance of the landraces that many farmers are still using, increasing the threat to the world's germplasm resources.)

The optimists, according to Bongaarts, see no problem providing food for 10 billion people. But to do so, said Bongaarts, would require more than a quadrupling of current food production. The fear is that this would transform the world into a "human feedlot," in the Ehrlichs' phrase, with a severe impact on the environment. The land area devoted to agriculture would have to increase by 50 percent.

Ehrlich and Gretchen Daily admitted that technological advances that are currently unforeseen can alter the picture, but they counseled prudence. "Technical progress will undoubtedly lead to efficiency improvements, resource substitutions and other innovations that are currently unimaginable," they wrote. "Nonetheless, the costs of planning development under incorrect assumptions are much higher with overestimates of such rates than with underestimates." It is good to hope for the best, but it is best to plan for the worst.[5]

Bongaarts carefully steered a middle course: "In reality, the future of global food production is neither as grim as the pessimists believe nor as rosy as the optimists claim. Whatever the outcome, the task ahead will be made more difficult if population growth rates cannot be reduced."[6]

POPULATION

Paul Ehrlich, with his wife, Anne, and various colleagues, has been one of the most visible advocates of population planning. His 1968 book *The Population Bomb,* coupled with his repeat appearances on the Johnny Carson show to discuss it (Ehrlich was on the show twenty-five times), put the population question on the front pages of American newspapers for the first time.

"We've come a long way" since 1968, Ehrlich said, in terms of public understanding of the population problem. But the growth in understanding has not yet done much to solve the problem.

"When *The Population Bomb* was written, I was accused of being much too pessimistic," Ehrlich said. "And look at the things we didn't know about. We didn't know about acid rain. The ozone hole hadn't appeared yet. Norman Myers hadn't done his work on what was happening to the tropical forests, so we didn't know about that." Global warming was only partly understood at the time, and even many climatologists discounted it. "Basically it was an amazingly optimistic book." The world's population when that book was written was about 3.5 billion.

Since then, he said, "we have added 2 billion people to the population, and we haven't added anything like the amount of food needed to improve the diet. Africa has been falling behind for a long time. Central America is starting to fall behind. Asia is doing better."

The problem is not with the food supply, which has been increasing, but with the failure of efforts to slow population growth, Ehrlich said. "One of the old stories you hear is that if you increase the food supply—and don't control the population—you just get more people living hungry. That has proven to be true."

Costa Rica is one example of where progress has been made. Population experts have long believed that one of the ways to get birthrates down is to give women access to birth control and provide better health care for them and their children. "A major way to get birthrates down is empowering women," Ehrlich said.

"Well, Costa Rica has done a lot of that. They're probably more literate than the United States, if you can believe the recent poll data. Contraceptives are about as accessible to Costa Rican women as they are to American women. And their birthrate did come down. Average family size came down from about seven to a little over three children. Which isn't far enough, but it happened quite rapidly."

Then it hit a snag. The birthrate has not dropped much below three children per family. Ehrlich would like to see it drop below two, and he is trying to find out why it hasn't.

"I think the studies have neglected men too much," he said. "It's not just the status of women; it's the attitudes of men. I'm totally in favor of empowering women, but I think we have to pay attention to men's attitudes, too, because I think they have enormous influence." Costa Rican women are still expected to be mothers above all else. The same thing is true in Kenya, Ehrlich said. There even totally Westernized women, such as flight attendants, still want large families, he said.

The inequitable distribution of food is another problem, as Cynthia Green, an international development consultant, has pointed out. "Increased food supplies do not necessarily translate into more people fed and better diets among the neediest because much of the additional food is consumed by middle-income consumers who use their increased spending power to improve their diets," she said. Population growth accounts for only about half of the increase in food demand; the rest comes from people who simply want more—and can afford it.

"If all food were equally shared and you didn't feed grain to animals, we'd have enough. That is true," Ehrlich said. (Feeding grain to animals to produce meat is far less efficient than feeding grain to people.) "With everybody eating a vegetarian diet you could feed six or seven billion people, ballpark. But then you have to ask yourself the question: Is it going to be easier to ask people to have fewer children, or is it going to be easier to ask them to share everything and become vegetarians?

"My personal view is it's demonstrably easier to get people to have fewer children than it is to get them to share equally. Who the hell shares equally or ever has shared equally? If people would all learn to become vegetarian saints, you could have more people without overstressing the planet than if they're going to be typical Hollywood moguls."

In 1990, the Ehrlichs followed *The Population Bomb* with *The Population Explosion.* They updated the arithmetic of the first book, and the numbers were not promising. It seems that no matter what steps are taken now, the world's population is almost certain to double. The reason is what the Ehrlichs call "demographic momentum."

In 1989, 40 percent of people living in developing countries were less than fifteen years old. More than 1 billion people in those countries have their reproductive years ahead of them. The teenagers and children in that bubble will have children and grandchildren before they contribute to the death rate.

If those children marry and have an average of about two children per family, population growth will stabilize. But that will not happen until the average age of the population rises enough for the death rate to keep pace with the birthrate. When the population, on average, is very young, there are far more children than deaths, even if the birthrate averages two children per family.

The time it takes for population to stabilize is about a lifetime—fifty or sixty years in the developing world. In India, to use Ehrlich's example, the population in 1990 was 850 million, and Indian families had an average of 4.3 children. The "replacement rate" in India is 2.4 children per family. (It is slightly higher than two because it takes into account the high infant mortality in India.)

Say Indian families reached the replacement level in a single generation—over the next thirty years. India's population would continue to grow for a century and reach 2 billion. "That's what demographic momentum is all about," the Ehrlichs said. Unless the high fertility rates in Asia and Africa decline during the 1990s, the world is almost certain to reach a population of 10 billion before it levels out. If the birthrate declines do not occur until the twenty-first century, the peak population will be even higher.

"That means we've got to put a lot of effort into increasing the food supply even at great ecological risk," Ehrlich said. "Otherwise huge numbers of people are going to starve to death."

"Back in 1950, most nations were self-sufficient, and I would say almost half of them, at least fifty to seventy-five, were exporting food," Pimentel said. "That's now down to about four or five nations that are exporting food. The U.S. is the largest exporter. Now our population is projected to go to five hundred million, half a billion, in sixty years. We won't have any food to export. All I can ask is: Is that what we want?"

The way to solve the population problem, Ehrlich said, is to limit births, not increase deaths. "It is my moral position that . . . you do everything you can to take care of the people you've got here now.

"I debated a demographer who said there have always been starving people and there always will be. He said it with a smile, and I've described him as being able to bear other people's suffering with a stiff upper lip."

Some have suggested that the AIDS epidemic now ravaging Africa will help take care of population problems there. Others suggest starvation is inevitable and likewise helps control population, Ehrlich said.

"It's perfectly true," he said. "There's no worry about the population explosion ending. It's just a question of how it happens."

SOIL

"The dust lifted up out of the fields and drove gray plumes into the air like sluggish smoke," John Steinbeck wrote in the *The Grapes of Wrath*. "The finest dust did not settle back to earth now, but disappeared into the darkening sky."[7]

It was the 1930s, and for the first time, American farmers were experiencing soil erosion on a massive scale. Thousands of them were forced off their land to begin the migration to California that Steinbeck described in his novel. The crisis was brought home to members of Congress in Washington in 1934. Leaving a legislative session late one afternoon, they found dust all over their cars. The dust came from Oklahoma and Kansas, the congressmen were told. They immediately took action to establish the Soil Conservation Service in the Department of Agriculture, which helped ease the crisis.

The problem of soil erosion did not go away, however. According to Lester R. Brown, the president of the Worldwatch Institute in Washington, the problem intensified in 1972 after the sale of wheat to the Soviet Union led to a doubling of grain prices. The amount of cultivated land in the United States grew by 10 percent as farmers covered every available space with crops to capitalize on the higher prices. The increase in cultivation led to a marked increase in soil erosion.[8]

Agriculture depends on soil, water, and energy. Of the three, Pimentel ranks soil as the most crucial resource. "I rank soil as number one because it takes so long to form soil," he said. "Under agricultural conditions, it takes five hundred years to get one inch of topsoil. Five hundred! You need a minimum of six inches for agricultural production. That's three thousand years. You can't sit around and wait for it."

Soil, Pimentel explained, is not the inert substance it seems to be. It is actually a complex tissue of microbes, insects, and other creatures. "It is really a living organism, in a sense, and essential to agriculture," he said.

Earthworms, for example, knead the soil, pulling wastes underground and leaving holes for water to fill, preventing water runoff and soil erosion. "Their numbers can be enormous," Pimentel said, up to nearly one-half ton per acre. An acre of land can also harbor more than a ton of bacteria and an even larger mass of fungi. "If we didn't have these organisms, the crop residues, the livestock wastes, and so forth, would tend to accumulate and not be recycled and the nutrients not be put back into the system," Pimentel explained.

Soil has become increasingly important as the demand for food has risen, Pimentel said, beginning in prehistoric times. "When humans first evolved on earth, we were hunter-gatherers. We lived off the land, primarily on vegetable-type material. Nuts and roots and seeds, and whatever we could find. To a small degree, of course, we were hunting for various animals. But we were also eating insects and frogs and lizards and snakes. They were easier to capture than a buffalo or whatever."

That kind of living off the land probably required five hundred acres per person, Pimentel said. As the population increased, that kind of land was no longer available. That marked the beginning of agriculture, when prehistoric human ancestors began planting seeds of foods they had found.

"We were concentrating our food production in that ecosystem, though we still were not managing it. We just scratched in the seeds and came back three or four months later and harvested what was there. What the insects and diseases and weeds and the wild animals didn't take, we got. But it was more than we were getting previously."

Next, the early farmers began to try to keep weeds out and protect the plants from birds and diseases—all of which increased the harvest. "And then the next stage was generally more intensive management, applying manure from our livestock that we now had for power and for milk and meat." The final stage was the use of fossil fuels to produce pesticides and fertilizers, pump irrigation water, and fuel tractors.

Expansion of the amount of land being farmed was one factor that helped boost food production in the twentieth century. Now most of the land on earth that is suitable for agriculture is already being farmed. There is no more. Between 1950 and 1981, the land planted with grain around the world increased 24 percent.[9]

"And now look what has happened, the way we're mismanaging our land," Pimentel said. Iowa reported in 1981 that it had lost half of the topsoil that was there before farming began about one hundred years ago. "That's some of the richest soil in the nation, or the world, for that matter," Pimentel said. A region in the state of Washington reported that it had lost 40 percent of its topsoil over the past century. Since 1981, the land planted with grain has fallen by 7 percent worldwide.[10]

About 2.5 million acres of U.S. cropland is abandoned each year because of soil degradation, Pimentel said. American fields are losing, on average, about seven tons of topsoil per acre each year. In Iowa, it is twelve tons per acre. Tennessee and Kentucky are nearly that high. An estimated 3 billion tons of topsoil are lost to watersheds in the continental United States each year.[11]

Before farming accelerated the losses, the amount of soil lost on each acre was probably about one twenty-fifth of a ton, Pimentel estimated. Two-thirds of the eroding soil is running off into lakes, oceans, and rivers, including the muddy Mississippi, he said, and it cannot be recovered.

Since 1972, deserts have expanded by almost 500 million acres worldwide. The world's farmers have lost 480 million tons of topsoil, more than all of the topsoil on all U.S. farmland.[12]

By 1977, American farmers were losing six tons of topsoil for every ton of grain they grew. Even where it did not render cropland useless, the loss of topsoil took a toll. Studies showed that for each inch of topsoil lost on cornfields, the productivity of those fields declined by 6 percent. The same was found to be true for wheat.[13]

Because of the long time it takes to make soil, only about one-half ton of newly minted topsoil is restored to the fields each year. Again, the living organisms in the soil—the soil "biota," as Pimentel calls it—are crucial to the production of new soil. They help recycle nutrients.

Dung beetles, for example, roll up a little ball of cattle dung, put an egg in it, and bury it. Australia learned about the importance of dung beetles when it started raising cattle and sheep. Australia has no native dung beetles. "The dung was accumulating on the surface of the land there because it didn't have the right insects to recycle it," Pimentel said. "The Australians were going to be buried in cow manure if they didn't have the right biota." The problem was solved by importing dung beetles from Africa.

There are about forty tons of organic material in an acre of soil, and it takes time to incorporate that, Pimentel said. Earthworms bring twenty to forty tons of soil to the surface each year. Insects move about one-tenth to one-fourth as much. "Ants, for example—you've seen the little piles of dirt with ants," Pimentel said.

Iowa is losing soil at a greater rate than other parts of the nation partly because of the intensive pressure of agriculture, with fields being planted year after year. In the 1940s it took two acres to grow an acre of corn. One would be planted in corn, the other in a legume, such as sweet clover, to cover the soil, slow erosion, and restore nitrogen. The next year, the two crops would be swapped. Because the legumes restored nitrogen to the soil, chemical fertilizers weren't needed.

"This is an insidious, nasty degradation of an essential resource for food production," Pimentel said. He has estimated that soil erosion and associated problems with water runoff cost the United States $44 billion per year.

WATER

Pimentel ranks water as the second most vital resource to agriculture, after soil. Modern agriculture's demand for water is enormous. "It takes about fifteen hundred pounds of water to produce one pound of corn," he said. "And about twice that if you're going to produce rice." Put another way, an acre of corn requires about five hundred thousand gallons of water if it is relying on rainfall. An acre of irrigated corn needs 1 million gallons of water during the course of the growing season.

Eighty-five percent of the water consumed in the United States is consumed by agriculture. "I underline *consume* because we take the water out, we use it, and it's released into the atmosphere," he said. "Humans, the public and industry, consume only 15 percent of the water. They use the water, and they put it back. Now they frequently put it back polluted, but at least they put it back. In agriculture, we use it, and we do not put it back. It's actually pumped through the crop and released into the atmosphere."

Much of it also disappears into the food itself. Seventy to 80 percent of fruits and vegetables is water, which is pumped onto the soil, taken up by the fruit, and shipped off the farm. Pimentel likes to use lettuce as an example. It is 95 percent water. During the winter, California water is shipped inside heads of lettuce, with a considerable expenditure of energy, to the East Coast.

A head of lettuce contains fifty calories of food energy, Pimentel said. It takes the equivalent of four hundred calories of energy to grow a head of lettuce and another eighteen hundred to ship it east. That is the equivalent of twenty-two hundred calories of energy expended to deliver fifty calories of food energy on the East Coast. And California has lost the water.

The problem of water loss is closely tied to the problem of soil erosion. Water is responsible for three-fourths of the erosion in the United States; wind is responsible for the rest. If the soil is protected, the water runoff slows, and the water is conserved.

Irrigation has been used to produce food since biblical times. Water from the Tigris and Euphrates rivers moistened the fields of the Mesopotamians. A system of tunnels called *kanats* carried water downhill from the mountains to farmland twenty-five hundred years ago in what is now Iran.

Irrigation has now become "a cornerstone of global food security," according to Sandra Postel of the Worldwatch Institute. The world's irrigated farmland is five times what it was in 1900. Seventeen percent of global farmland is irrigated, and it produces one-third of the world's harvest, she said.[14]

The amount of irrigated farmland in the United States rose for several decades before 1978. Then it began to decline. Since 1978, the amount of irrigated farmland in the United States has dropped by 7 percent.[15] Around the world, the amount of irrigated farmland grew by 2–4 percent per year during the 1960s and 1970s. Increases slowed during the 1980s. One reason is that irrigation projects are becoming more expensive. Deeply indebted developing countries cannot afford them.

Although it is a cornerstone of agriculture, irrigation has taken a toll on the environment. Most water contains at least small amounts of salt. When huge amounts of water are poured over the soil, the salt builds up in the soil, eventually making it too salty to support crops. The Tigris and Euphrates river basins are now severely affected by salt buildup.

Crop plants have more difficulty surviving in salty soil. In the United States, crop yields are dropping on more than one-fourth of the nation's irrigated acres because of this salt buildup.[16]

Water also picks up contaminants in the soil as it percolates through it. Selenium is one that has become a dangerous contaminant in the western United States. "Irrigation has washed more selenium and other dangerous chemicals out of the soil in several decades than natural rainfall would have done in centuries," Postel said.

Selenium is toxic at high levels, and the selenium-enriched water running off California farms is killing birds and fish, and it could pose a threat to humans as it moves up the food chain. It has now been found at hazardous levels in twenty-two wildlife areas in the western United States.[17]

What is most important to the coming squeeze in agriculture is that, as Pimentel noted, water is simply being used up. "We're pumping out our groundwater resources, our aquifers, 25 percent faster than our recharge rate," he said. As the level of the aquifers drops, it requires more energy to pump the water out of those aquifers. Instead of pumping water from 100 feet below the surface, say, it becomes necessary to pump it from a depth of 150 feet. And that can never be corrected.

"No matter how much rain you allow for the recharge to stabilize, you're always going to have to pump 150 feet, because you don't refloat the soil again," Pimentel said. "That means now you've got to continue to pump that from here to eternity. And of course you can pump them out, too—totally pump them out." As the underground aquifers are drained, the land subsides. In some parts of California, the surface has dropped twenty-nine feet, Pimentel said.

The Ogallala aquifer, a huge underground storehouse of water that reaches from South Dakota to Texas, is a critical source of irrigation water in the western United States. Like many other aquifers, it is being emptied faster than rainfall can recharge it.

In the continental United States, water is being used about 25 percent faster than it can be replenished. In Texas, that overdraft reaches 77 percent, according to Pimentel. Much of that water is wasted, he said, because of government water subsidies that bring the cost of water down. In Utah, for example, farmers pay about eighteen dollars per acre for water that the government subsidizes at the rate of about six hundred dollars per acre.[18]

According to one calculation, a 20 percent increase in the world's population will double the demand for water. "The prospect for future expansion of irrigation to increase food supplies, worldwide and in the U.S., is not encouraging," wrote Pimentel and Henry Kendall, a Nobel Prize–winning physicist from MIT. "Greatly expanded irrigation is a difficult—if not an unsustainable—solution to expansion of agriculture."[19]

ENERGY

Americans consume the equivalent of about four hundred gallons per person each year in the production of food. That's about 17 percent of America's energy. That includes all the fossil fuels used for irrigation, to produce pesticides and fertilizers, and to provide gasoline and diesel fuel for farm equipment, according to Pimentel. But that is only part of what is involved in bringing food to the table. Growing the food consumes one-third of all the energy used for agriculture.

"You've got to process and package, and you've got another third of the total quantity of energy involved there," Pimentel said. "And then distribution and cooking account for about another third."

American agriculture uses far more energy than, say, Chinese agriculture, Pimentel said. The Chinese expend five hundred hours of labor for each acre of corn they grow. In the United States, an acre of corn requires four hours of labor. The Chinese system is good for China, where there is adequate manpower to supply the labor.

"That's the one surplus they've got—manpower," said Pimentel. "They wouldn't know what to do with all those people. So this is a place that, tentatively at least, the manpower is paying off. They still use as much fertilizer as we use, and about as much irrigation."

The American system reduces the labor by relying far more than the Chinese on external energy, especially fossil fuels. Even so, the Chinese are using one hundred times as much energy for farming as they did in 1955, Pimentel has calculated. In the United States, the use of fossil fuels in various parts of the economy is now twenty to one thousand times what it was thirty or forty years ago.

Each American, on the average, uses as much energy as 2 Swedes, 3 Greeks, 33 Indians, or 295 Tanzanians.[20]

PESTICIDES AND FERTILIZERS

In 1992, researchers at the National Cancer Institute reviewed about two dozen studies of the health of farmers and came to a disturbing conclusion. The health of the farmers was generally better than their urban counterparts. They had lower rates of heart disease and lung cancer, probably explained by their low rate of cigarette smoking and high level of exercise, said Aaron Blair of the institute.

What was striking was that the farmers had much higher rates of other cancers, including Hodgkin's disease, multiple myeloma, leukemia, melanoma, and cancers of the lip, stomach, and prostate. The melanoma and lip cancers are probably due to farmers' exposure to ultraviolet radiation in sunlight, Blair said. But the other cancers could be due to exposure to pesticides.

"These aren't hard links, you understand, but it's more of a suggestion of one of the things that might be going on," Blair said. "Pesticides affect the immune system in a number of animals." Pesticides are also increasingly spreading to urban and suburban areas, perhaps explaining the more recent rise in some of these same cancers in nonfarmers. Other possible causes of the cancers are fertilizer in drinking water; fuels and oils; fumigants; and animal cancer viruses, Blair said.

"This is saying there is a weight of evidence, there is a growing body of scientific studies, and they all point in the same direction—that there are troubling increases in certain kinds of cancer in farmers that are also increasing in the general population," said Devra Lee Davis, a scientist in the U.S. Health and Human Services Department and an authority on environmental causes of cancer. "It's not that we should have no chemicals for agriculture but that we need to be more prudent in using the ones we have."

In April 1993, the cancer institute study was followed by another piece of research taking a new look at DDT. Mary S. Wolff, a chemist at the Mount Sinai School of Medicine in New York City, looked for evidence of DDT exposure in women with breast cancer. She found that the women with the highest exposure to DDT had four times the breast-cancer risk of women with the lowest exposure.

While the findings do not constitute proof that DDT causes breast cancer, they could, if confirmed, provide a possible explanation for the puzzling rise in breast cancer in recent decades in the United States. "Breast cancer is the most common cancer among women, and a lot of the risk is unexplained," Wolff said.

The rise in breast cancer followed the increase in the use of DDT. Even though DDT was phased out in 1972 in the United States, Americans are still exposed to it through their diets. Before 1972, it was common in meat and dairy products. Because it is stored in the body for decades, most Americans still carry DDT residues, Wolff said. Children are exposed to it through their mothers' milk. And it is still widely used in other countries.

DDT levels in Great Lakes fish, which dropped sharply after the 1972 phaseout, are now increasing again, probably because DDT used elsewhere is being picked up by atmospheric currents and deposited in the lakes, she said.

Wolff and her colleagues measured levels of a DDT-breakdown product in the blood of 58 women with breast cancer and 171 women without breast cancer. Women with levels in the top 10 percent had four times the breast-cancer risk of women in the bottom 10 percent. The researchers also looked for a link between PCBs and breast cancer but failed to find one. PCBs, or polychlorinated biphenyls, are hazardous liquids used as insulators in electrical transformers. Like DDT, they are widespread environmental contaminants.

In a commentary on the study, David J. Hunter and Karl T. Kelsey of the Harvard School of Public Health in Boston said: "Because the findings . . . may have extraordinary global implications for the prevention of breast cancer, their study should serve as a wake-up call for further urgent research."

Pesticides and fertilizers have made important contributions to the rise in agricultural yields since World War II. Growing pressure among consumers and some scientists to restrict pesticide use is again squeezing agriculture, limiting farmers' options. As with the other factors already mentioned, limits on pesticides will increase the demands on plant breeders to produce better crop varieties—specifically, varieties resistant to more diseases and pests.

Many consumers object to what they believe is harmful and often needless contamination of their food with agricultural chemicals. A dramatic example of that anger concerned the use of Alar, a growth regulator used to improve color and storage in apples.

In February 1989, the Natural Resources Defense Council published a report linking Alar to cancer. The report's findings were widely disseminated, thanks to a skillful public relations campaign that included disclosure of the report on CBS's *60 Minutes.* Meryl Streep lent her name to a campaign to ban Alar. Mothers owed it to their children, she said, to see that chemicals were eliminated from apples.

The campaign scored a direct hit. Consumers responded immediately with demands for Alar-free apples. Apple growers panicked. The Environmental Protection Agency (EPA) moved quickly to announce that it would ban the use of Alar. Before the agency could do so, Alar's manufacturer, the Uniroyal Chemical Company, withdrew Alar from the market.

Consumer concern over agricultural chemicals can exert a powerful effect on agricultural practices. Some farmers may share consumer concern over the health and environmental threats posed by pesticides. They also have a special reason for wanting to do away with them or at least reduce their reliance on them.

One reason is that farmers are often exposed to far higher levels of pesticides than are consumers. Careful techniques are required to reduce farmers' exposure during pesticide application. Farmers and their children are at most risk of exposure to the pesticides once the chemicals have been deposited on the fields. Any potential health risks associated with pesticides are multiplied for farmers and their families.

Another reason farmers are reluctant to use chemicals is that they are expensive. Fertilizer use increased dramatically after World War II. In 1984, farmers were using four times the fertilizer they had used in 1950, according to Lester Brown of the Worldwatch Institute. Since 1981, however, fertilizer use has declined in the United States. There was no shortage of fertilizer; it simply became too expensive. And it was not generating the increases it had in the past. Twenty years ago, each additional ton of fertilizer applied in the U.S. Corn Belt increased the harvest by fifteen to twenty tons. Now a ton of fertilizer boosts yields by only five to ten tons.[21]

Pesticides and fertilizers can also have a bruising effect on the environment. They accumulate in surface water and groundwater and can spread quickly over a large area.

In a 1990 report, the EPA found that agricultural runoff—the soup of chemicals washed off farmland by the rain—is the leading cause of pollution in the nation's rivers. In the EPA's survey, agricultural runoff was found to have some effect on 55 percent of the polluted portions of American rivers.[22]

Agricultural runoff doesn't emerge from an easily identifiable point source, such as an industrial-waste pipe. Industrial waste can be kept out of waterways by capping waste pipes. Agricultural runoff cannot be turned off. The only way to reduce it is to reduce the use of pesticides and fertilizers.

Pimentel began his scientific career with a study of pesticides. For his doctoral thesis at Cornell, he looked at insecticide resistance in houseflies. He looked for a better way to control them rather than simply upping the dose of insecticide. "I was able to devise a scheme using one-thousandth the quantity of insecticide in barns," he said. "And it provided more effective control of the flies. So I was off and running."

Pimentel then left Cornell to join the Public Health Service and pursue an interest in veterinary medicine. He was sent to Puerto Rico, where he worked on snails that carried the parasites responsible for the human intestinal disease schistosomiasis. He also looked at the habits of the mongoose, which was a primary carrier of rabies.

"The mongoose in Puerto Rico was introduced intentionally for rat control," Pimentel recalled. "What happened was it controlled one rat that was in sugarcane, the Norway rat. But when that rat was reduced, it reduced the competition for the tree rat, which the mongoose could not deal with—the mongoose does not climb trees. So all they did was change rat species with the introduction of the mongoose. And now they've got the mongoose that kills nesting birds, and of course it's a vector and reservoir for rabies. It's just a disaster. That's one of the classic examples of biological control that turned out to be a disaster."

Pimentel left the Public Health Service for a fellowship at Oxford University, then returned to Cornell. During the Johnson administration he was asked to serve on a presidential commission on restoring the environment. He was chosen for his expertise on pesticides. His service on the commission helped establish him as a leader in the study of the effects of pesticides on the environment.

Pimentel is particularly concerned about the harmful effects of pesticides on agriculture. The reason is that pesticides kill beneficial organisms as well as pests. The underground ecosystem of earthworms and microbes that make soil suitable for farming are especially vulnerable to pesticides, Pimentel said. So are many of pests' natural enemies. Often, when the pests are killed, their natural enemies are killed, too, allowing them to spring back faster.

Honeybees are another example of beneficial insects that have been devastated by pesticides, Pimentel said. "About $30 billion worth of crops in the United States are pollinated by honeybees and wild bees," he said. In some areas, pesticides have wiped out so many bees that the crops are not pollinated.

"There's a new business that's gotten started by apiculturalists who now rent bees to orchardists and melon growers and so forth so they can pollinate their crops. This is due to the destruction of bees by pesticides," Pimentel said.

American farmers spend about $4 billion a year on pesticides to protect crops from insects, diseases, and weeds, Pimentel said. Even so, pests destroy about 37 percent of all potential food production on the nation's farms. But that $4 billion in pesticides is probably saving about $16 billion in crops that would otherwise be lost. Without including the environmental and public health costs associated with pesticides, that is a four-to-one return on the investment. Even when environmental and health costs are figured in, pesticides are still profitable, Pimentel said.

Then he ticked off some of the environmental and health consequences. They include human pesticide poisonings and deaths, cancer, adverse effects on the immune system—and sterilization. "Of course, we need more of that," he said, laughing. "But not that way."

Among the other adverse consequences are the poisoning of domestic animals, including cats, dogs, and cattle. Fish and bird kills are common, as well as the destruction of pests' natural enemies. "We estimate that about half a billion dollars is spent in additional pesticides to deal with the pests that are created because we destroyed these natural enemies," he said.

Pesticides also destroy crops. "Some of these herbicides, in particular, drift when they're being applied to one crop and destroy other crops," Pimentel said. In one example, farmers in Texas used airplanes to apply a herbicide to control weeds in wheat. The herbicide drifted onto cotton in neighboring fields, destroying $20 million worth of it. "This happens with grapes, which are also very sensitive, and with tomatoes," Pimentel said.

"Another impact you can get from herbicides is it makes the corn crop more susceptible to insects and diseases." In research done by Pimentel and his colleagues at Cornell, the corn-leaf aphid was three times more abundant on corn that had been exposed to a herbicide than on corn that had not. "Corn borers were more abundant, and they grew larger and produced more eggs," he said. "The corn was then no longer resistant to another corn-leaf blight. And black corn smut was five times larger" on the corn exposed to the herbicide.

Pimentel does not know exactly why that happens, but it is clear that the herbicide somehow changes the biochemistry of the corn. In corn exposed to the herbicide, for example, the nitrogen content was higher. "We believe that the high nitrogen content made the corn more nutritious for the insects and diseases," he said. "I feel sure it was playing a role, but we don't know how important."

Herbicides can also increase the natural toxins in some crops, he said. "It's been proven that cattle have been killed and sheep have been killed using herbicides on some of our different crops," Pimentel said.

The increasing use of pesticides may have boosted yields and farmers' profits, but it has not stopped insects. Corn, for example, is now the largest user of insecticides in the nation. It surpassed cotton a few years ago. In 1945, before insecticides were widely used, about 3.5 percent of the corn crop was lost to insects. It is now 12 percent. During that period, the use of insecticides increased a thousandfold or more, Pimentel said.

Much the same thing happened with other crops. Insecticide use on all U.S. crops has increased tenfold since 1945. At the same time, crop losses have nearly doubled, from 7 percent to 13 percent.

Changes in the way crops are grown made them more vulnerable to insects. In 1945, for example, corn was grown in rotation with soybeans, wheat, and other crops. When insecticides came along, farmers realized they could "grow corn on corn" rather than rotating crops. "When you grow corn on corn, you end up with a very serious insect problem," Pimentel said. "You also end up with disease and weed problems."

The change in farming practices also aggravates many of the other environmental problems associated with modern agriculture. "You increase water runoff when you grow corn on corn, you increase soil erosion, and you have greater contamination of the soil with pesticides." Along with the soil and water that's lost, $18 billion in fertilizer is being washed away each year, Pimentel said.[23]

GLOBAL WARMING

The year is 2050. Because of the greenhouse effect, the earth's average temperature has risen five degrees. The earth is hotter than at anytime in the past 4 million years. The Great Plains have been a desert of shifting sand dunes since a new wind pattern scattered the topsoil eastward. U.S. grain reserves have periodically dropped to zero, but grain has been imported from Canada, which has cut its forests to expand its grain acreage.

As the climate warmed, eastern forests succumbed to a line of death marching northward. Only dead skeleton trees remain. Wetlands dried up, and migratory birds vanished, unable to survive journeys across a parched continent. The village of Churchill on Canada's Hudson Bay has boomed, as have other Arctic towns. Many of the world's coastal areas have been inundated by rising seas. Dikes have been built around Boston and Charleston, and they are planned for New York City. Temperatures are predicted to continue rising.[24]

This scenario comes from *Dead Heat* by Michael Oppenheimer, a scientist at the Environmental Defense Fund, and the writer Robert H. Boyle. The book opens with a grim but realistic picture of what might happen as a result of the phenomenon known as global warming, or the greenhouse effect.

Researchers disagree as to how warm the earth will become and how fast the temperature rise will occur. But many believe the consequences could be quite severe. Some farmers could be forced to abandon their farms, reminiscent of the Dust Bowl years of the 1930s. The American breadbasket could migrate northward into Canada.

Although researchers disagree over predictions of the amount of global warming, they all agree on one central, observable fact: The carbon dioxide content of the atmosphere is climbing. Continued reliance on fossil fuels and the burning of tropical forests are pumping 5 billion tons of carbon into the air each year in the form of carbon dioxide.[25]

During the hot, dry summer of 1988, talk of global warming was on everyone's parched lips. Because of the uncertainty over the extent and pace of global warming, scientists did not know for sure whether the drought of 1988 was partly due to rising carbon dioxide levels. But that was the kind of summer that many of them expect to become more common as global warming continues. Whether or not the summer of 1988 was itself a consequence of global warming, it was a taste of things to come.

The summer of 1988 also demonstrated what global warming could mean for agriculture. For the first time in history, the United States did not produce enough grain to feed itself. The corn crop alone declined by 40 percent in 1988.[26] Some authorities worry that new crop varieties cannot be developed quickly enough to cope with rapid changes in the environment. That is almost certain to lead to a drop in food production. In a worst-case analysis, global warming could lead to crises twice per decade, each resulting in the starvation of 50–400 million people.[27]

Global warming is one of several environmental threats that represent something new in human history. For the first time, human beings have acquired the ability to modify the environment on a global scale. Other global environmental problems include the thinning of the earth's protective ozone layer and the widespread pollution that has given rise to acid rain. By altering the environment in unpredictable ways, each of these problems threatens agriculture.

New crop varieties will be needed to survive possibly hotter and drier growing conditions in temperate climates, for example. Farmers will need crops resistant to acid rain and to the increased ultraviolet radiation reaching the earth's surface as a consequence of the thinning of the ozone layer. These pressures on agriculture will increase the demand for genetic resources to breed new crop varieties. Researchers will be seeking genes that confer drought resistance or make plants more tolerant of acid rain or ultraviolet radiation.

The same conditions that will demand more of plant breeders will also, however, accelerate the destruction of plant genetic resources. The climate changes that threaten agriculture will also threaten in situ reserves. Breeders will be caught in a squeeze, facing a greater demand for genetic resources that are disappearing. As another National Academy of Sciences panel noted, "Any future change in climate is likely to increase the rate of loss of biodiversity, while it increases the value of genetic resources."[28]

Agriculture, because of its obvious sensitivity to climate change, will be affected by even the most modest changes in the earth's temperatures. Because the precise consequences of global warming are difficult to predict, the future of agriculture can be assured only by making the system more flexible and adaptable to change. The development of genetically diverse crop varieties able to withstand a wide variety of environmental stresses can provide that flexibility.

In September 1991, a panel of distinguished scientists assembled by the National Academy of Sciences issued a report on the problems posed by global warming. The report, and the reaction to it, highlighted the uncertainty surrounding the potential consequences of global warming. The report painted a far more benign picture of its consequences than did Oppenheimer and Boyle in their future scenario. The panel's surprising conclusion was that adapting to climate change would be relatively painless.

Before it was even published, however, the report drew sharp criticism from a member of the panel that prepared it. Jane Lubchenco, an ecologist at Oregon State University, disavowed the report in a dissent that was included in the report. "I disagree with the report's implicit message that we can adapt with little or no problem," she said.

The report reached that conclusion by examining the effect of global warming on different human activities. For example, the report determined that farming would be sensitive to climate change but that farmers could adapt "at a cost." Natural areas and marine ecosystems were determined to be at great risk because they cannot easily be modified to withstand warmer climates. But virtually everything else the panel addressed—from industry, health, and tourism to water resources and coastal settlements—could be adapted "at a cost," the panel concluded.

Lubchenco charged that the panel grossly oversimplified the analysis by not looking at the environmental consequences of the adaptations it recommended. She said the panel paid too little attention to the difficulties of preserving natural areas and marine environments in the face of global warming. "The areas in which adaptations cannot be made, or at least not easily made . . . are so fundamentally important to the global system that mitigation—not adaptation—becomes paramount," she said.[29]

Stephen Schneider of the National Center for Atmospheric Research, an authority on global warming, said he objected to the panel's "surprise-free scenario of mild, predictable change."

Schneider and others believe that although the warming of the planet will be gradual, local climate change is likely to be abrupt and unpredictable.

William Nordhaus, a Yale University economist and a member of the panel, defended the report. He argued that even substantial losses in agricultural productivity would have little effect on the economy. "Agriculture, the part of the economy that is sensitive to climate change, accounts for just 3 percent of national output," Nordhaus told *Science* magazine after the report was released. "That means there is no way to get a very large effect on the U.S. economy. It is hard to say it is the nation's number-one problem."

Nordhaus's remark drew a response from Herman E. Daly, an environmental economist at the World Bank, who noted that "the current 3 percent figure could soar to 90 percent in the event of a serious disruption of agriculture." Even as Nordhaus defended the report, however, he admitted that it "is very much at variance with the scientific consensus you hear at meetings."[30]

In a section dealing with farming, the report said that increasing use of irrigation and the introduction of new crop varieties "likely can keep ahead of the climate change." It noted frightening examples of the effect of climate change on agriculture, such as those that occurred during the drought of 1988.

More frequent and more severe droughts are one possible consequence of global warming. Although the report acknowledged the severe consequences of such droughts, it maintained that farmers could adapt to them, primarily by substituting hardier new crops and increasing their reliance on irrigation. In the future, the report said, adaptation will become even easier with progress in biotechnology and computer-guided irrigation. But in order to adapt, farmers will need, among other things, "ample biological diversity in breeding," the report said.

But what of the effect of global warming on the maintenance of biological diversity? The report was far less sanguine about that. "For the sort of climate change we are assuming, timely adaptation of every species and conservation of the countless cooperators in the natural landscape are highly unlikely," the report's authors wrote. It noted that biological diversity could be preserved in seed banks and botanical gardens, but it acknowledged the severe deficiencies in many of the world's seed banks. "*In situ* preservation of species and systems must carry the heaviest burden in programs to adapt to climate change," the report said.

The report's solution to the preservation of in situ reserves in the face of changing climate is to pick them up and move them. The impracticality of this solution was not lost on the report's authors. "The cost of regenerating a simple plantation of trees has already been shown to be high, and moving a whole system of plants and animals would be even more costly," the report said. The alternative is to establish "a system of corridors connecting natural ecosystems across latitudes and longitudes," the report said.

Given the difficulty of preserving the wilderness areas that already exist, the prospects for creating broad corridors along which ecosystems could migrate toward the poles as climate warms seem dubious. Even if such corridors could be established, however, many species would not be able to take advantage of them. Plant communities are able to migrate with shifts in climate. But many species are unable to move fast enough to keep up with predicted rates of climate change.

In a summary, the report's authors concluded that the ability of natural areas to adapt to climate change is "questionable." The report's rosy assessment of the outlook for agriculture was predicated on the survival of "ample biological diversity." If the outlook for maintaining biological diversity is questionable, the outlook for agriculture cannot be as assured as the report suggested.

The term "global warming" refers to a rise in the earth's average temperatures caused by the buildup of so-called greenhouse gases in the atmosphere. The greenhouse gases include water vapor, carbon dioxide, methane, and CFCs. These gases, and a few others, trap the sun's heat in the atmosphere. Solar radiation entering the atmosphere passes through these gases more easily than the infrared radiation reflected back from the earth's surface.

The principle is identical to that in a greenhouse or in a car left in the sun with the windows up. Heat gets in but cannot get out.

The blistering-hot steering wheel in a parked car in summer is a good demonstration of how much energy is involved. If a single automobile can trap that much of the sun's heat, imagine how much heat the earth's atmosphere can contain.

Human activity has led to sharp increases in atmospheric levels of carbon dioxide and methane. Carbon dioxide is produced primarily by the burning of fossil fuels and the destruction of tropical forests. Methane leaks from natural gas wells and coal mines. It is a product of decay and is produced in large quantities by rice paddies and cattle.

CFCs did not exist in the atmosphere until they were developed several decades ago. They are now used around the world as refrigerants and aerosol propellants. (In the United States, they are no longer used in aerosol cans.) Deforestation has also contributed to the greenhouse effect. Whether trees are burned, turned into lumber, or allowed to rot, the carbon they contain eventually enters the atmosphere as carbon dioxide. (Conversely, living plants extract carbon dioxide from the atmosphere for the carbon, their principal ingredient.)

The earth's average temperature is about sixty degrees. If there were no greenhouse gases in the atmosphere, the earth's temperature would be close to zero degrees. It would be a frozen wasteland unable to support human life. But too much of a good thing could prove disastrous. No one can be sure whether the earth's temperature has risen since the start of the Industrial Revolution and the large-scale discharge of carbon dioxide into the air. But Charles Keeling, a chemist at the University of California, San Diego, has collected thorough evidence on the rise in carbon dioxide levels.

In 1958, Keeling set up a carbon dioxide monitor on the slopes of Mauna Loa in Hawaii. He measured the concentration of carbon dioxide in the atmosphere at 315 parts per million. Keeling continued to monitor carbon dioxide levels regularly, establishing beyond a doubt that they were climbing. By 1990, the concentration had risen to 353 parts per million, a 12 percent rise in just over thirty years. Although measures of carbon dioxide levels a century ago at the start of the Industrial Revolution do not exist, researchers have estimated that levels increased by 25 percent since 1750.[31]

Most researchers assume that the rise in carbon dioxide will ultimately lead to a rise in the earth's average temperature. Many variables remain to be worked out, such as the effect of carbon dioxide on cloud formation and the role of the oceans in the global carbon cycle. Those unknowns make it difficult to predict how fast global warming will occur.

Several research teams have attempted to answer the question by developing complex computer programs that serve as models of the earth's climate. The models suggest that a doubling of preindustrial levels of carbon dioxide could raise the earth's average temperature by 3.4 degrees to 9.4 degrees. "The larger of these temperature increases would mean a climate warmer than any in human history," said the National Academy of Sciences. "The consequences of this amount of warming are unknown and could include extremely unpleasant surprises."[32]

A doubling of carbon dioxide levels could occur sometime during the next century. The exact time depends on the rate of fossil-fuel use. The more fuel burned, the quicker the doubling time. These rises in temperature could produce a wide variety of climate changes. But what would they do to agriculture? Was the National Academy of Sciences panel correct in assuming that farmers could easily adapt?

Stephen Schneider does not think so. In his book *Global Warming,* he said that even modest declines in agriculture could exact a stiff price. Suppose that climate change reduces the yields of U.S. farmers by only 10 percent. The United States would still be able to feed itself and would still have a grain surplus. "But it could lose billions of dollars in increased production costs and decreased return on investments to farmers, since they would have to maintain the same investments in the face of lower yields," Schneider said.

A number of studies, Schneider said, have suggested that global warming could lead to a warmer and drier climate in the U.S. Corn Belt and the plains states. Cultivation patterns would shift, improving yields in some areas and diminishing them in others. There would be winners and losers. This "redistribution of comparative advantage could have major international security and economic implications."[33]

Increasing carbon dioxide levels would have one beneficial effect on plants. Experiments have shown that plants grown in air enriched with carbon dioxide grow bigger and faster. EPA studies have shown that the effects of this carbon dioxide "fertilizer" are not likely to compensate for changes in climate. Again, predictions suggest that there would be winners and losers.

Fakhri A. Bazzaz of Harvard University maintains a small cluster of greenhouses on the Harvard campus. Each is isolated from the others, and the air inside them contains differing amounts of carbon dioxide. After years of growing plants inside those greenhouses and comparing the results, he has shown that different species of plants respond differently to increasing levels of carbon dioxide. Rising levels of carbon dioxide—even if they do not produce substantial temperature increases—could have a profound effect on agriculture, Bazzaz has found.

Certain crops, such as soybeans, might be able to compete better against weeds in a carbon dioxide–rich environment. But other crops, such as corn and sugarcane, would fare poorly in the competition with weeds as carbon dioxide levels rise. The increased competition from weeds could cause yields to drop, Bazzaz said.

He also concluded that rising carbon dioxide levels are likely to wipe out many endangered plant species, further eroding the world's genetic diversity and further limiting plant breeders' options.

"It is clear that high carbon dioxide levels will have wide-ranging consequences for the natural world," Bazzaz said. "And it is clear that the carbon dioxide fertilization effect does not guarantee a lush, green future of agricultural abundance. . . . Such an atmosphere will not help lessen the planet's environmental and demographic woes. This atmosphere may induce climatic modifications that could undermine the integrity of the biological systems on which all *Homo sapiens* depend."[34]

The critical point is that rising levels of carbon dioxide—

which are a measurable fact, not a prediction—can affect plants dramatically even without any rise in global temperatures. If the dire predictions of global warming are wrong, climate may not change dramatically. But agriculture is certain to change.

Corn's growth, to take one example, is unlikely to be affected by rising carbon dioxide levels, Bazzaz's experiments suggest. But corn grown in atmospheres with elevated carbon dioxide contains less nitrogen. "That could mean lower protein," Bazzaz said. "Nitrogen is the major constituent of protein." His experiments have also shown that plants grown in carbon dioxide–rich air put more of their growth into their roots. "It's possible that could reduce the yield" of important crops, he said.

If global warming does occur, it is likely to increase the need for irrigation. Schneider describes a study by Dean Peterson of Utah State University that found that with a temperature rise of slightly over five degrees, U.S. irrigation needs in various locales would increase by 15 percent—if rainfall remained unchanged. If precipitation dropped by 10 percent, the need for irrigation would climb by 26 percent, Peterson concluded.

Higher temperatures could also affect grain quality, Australian researchers have suggested. They have found that Australian prime hard wheat makes poorer dough in warmer years. Experiments showed that wheat produces certain proteins in response to heat that can weaken dough. Even if global warming does not affect wheat yields, it could affect the quality of the crop, driving prices down and putting farmers out of business.[35]

Efforts to assess the effect of global warming on agriculture will remain difficult until predictions about temperature rise and climatic consequences become less uncertain. The computer models used to forecast global temperature rises cannot predict with any certainty the effect of warming on a particular region. But most researchers say uncertainty is no reason for complacency. On the contrary, as Schneider said, the uncertainty is worrisome. "To me," he wrote, "these divergent opinions and vastly differing forecasts of agricultural benefits and catastrophes suggest that, in agricultural terms, rapid climate change is a global gamble."

OZONE

In 1985, British researchers making atmospheric measurements in Antarctica made a surprising discovery. They found apparent damage to the ozone layer in the stratosphere over Antarctica. What had caused the ozone "hole" was not known, but the implications were frightening. The ozone layer that wraps the earth is a kind of global sunscreen. Drifting mostly at altitudes between seven and fifteen miles, ozone molecules absorb much of the sun's harmful ultraviolet radiation. At a minimum, damage to that global sunscreen could lead to an epidemic of skin cancer. If the ozone layer were to be substantially destroyed, life might no longer be possible.

One form of ultraviolet radiation—called ultraviolet-B, or UV-B, radiation—is particularly dangerous to life. It can produce mutations in DNA, the substance that genes are made of. It leads to cataract formation in the eyes and affects the human immune system. Each 5 percent decrease in the ozone layer allows 10 percent more UV-B radiation to reach the earth's surface. In terms of human effects, a 5 percent decrease translates to an additional twenty thousand cases of skin cancer per year.

The recognition of the impact of ozone depletion on human health has become widespread. Yet the dangers to agriculture are still not widely recognized. Research has shown conclusively that UV-B inhibits photosynthesis. Increases in UV-B exposure can damage crops. And UV-B can damage wild-crop relatives, posing a threat to in situ reserves.

As in the case of global warming, a new environmental hazard threatens both agriculture and the genetic resources needed to rescue it.

The discovery of the Antarctic ozone hole was reported in 1985 by Joseph Farman and colleagues at the British Antarctic Survey. Farman observed the decline for several years, checking and rechecking his data before publishing it in the British scientific journal *Nature.*

The finding was met with some skepticism. Satellites able to monitor ozone levels had detected no such ozone decline. When data from NASA satellites were subjected to closer scrutiny, however, Farman's finding was confirmed. Although it had earlier been missed, the evidence for the ozone hole was there in the satellite data.

International scientific teams visited Antarctica in 1986 and again in 1987 to observe the thinning of the ozone hole. By the spring of 1987, the amount of ozone over the South Pole was down to half its normal level. Some researchers dismissed the phenomenon as a seasonal occurrence attributable to polar winds and unusual—but natural—alterations of the atmosphere's chemistry.

Soon, however, reanalysis of global atmospheric measurements in light of Farman's discovery showed that stratospheric ozone was being depleted around the world. Scientists found that the concentration of ozone in the stratosphere had declined by about 2 percent between 1969 and 1986. In the United States and Europe, stratospheric ozone levels were down about 3 percent all year-round.

Chemists had theorized that CFCs, which contribute to the greenhouse effect, were the principal culprits in the destruction of the ozone layer. Ozone is a molecule made up of three atoms of oxygen. (Oxygen in the air occurs as a molecule of two atoms of oxygen.) Ozone is formed when an oxygen molecule absorbs shortwave radiation. The radiation splits the molecule into two oxygen atoms, each of which combines with an oxygen molecule to form an ozone molecule.

Chlorine reverses that process. Chlorofluorocarbons, as their name suggests, contain chlorine. They were hailed as a great advance when they were discovered because they were chemically stable. That meant they were unlikely to react with living cells and therefore would not be toxic.

That same stability, however, is what makes CFCs damaging to the ozone layer. They do not break down in the lower atmosphere where they are released. Over a period of time, these stable molecules drift upward into the stratosphere. There they are broken down by the higher levels of ultraviolet radiation. Chlorine atoms are freed, and they act as catalysts for the conversion of ozone to oxygen. They speed the conversion but are not consumed by it. So each chlorine atom can convert thousands of molecules of ozone to oxygen. It is this multiplier effect that makes chlorine such a potent destroyer of ozone.

For a time, some chemists argued that natural processes were responsible for the Antarctic ozone hole and the destruction of the ozone layer. Experiments eventually convinced most researchers that CFCs were primarily responsible. CFCs were widely adopted as wonder chemicals because of their chemical stability. They did not react with other chemicals, and therefore they were safe.

But that stability allows the chemicals to reach the stratosphere, where they do their damage. Colder temperatures can facilitate the destruction of ozone, which probably explains why the worst damage occurs over Antarctica and to a somewhat lesser extent in the Arctic.

The United States and several other countries had banned the use of CFCs in most aerosol cans by 1978, but CFCs were still in widespread use when the Antarctic ozone hole was discovered. The Reagan administration was dismissive of renewed efforts to limit the use of CFCs. In 1987, Interior Secretary Donald Hodel suggested that the problem of the ozone layer could be taken care of by asking people to wear hats and sunglasses. Several environmental groups donned hats, sunglasses, and white sunscreen cream for a press conference at which they demanded Hodel's resignation.

"I don't believe we ought to box in the president," Hodel said. "We ought to provide an array of options." Hodel's hat-and-sunglass strategy, whatever it might have done to cut skin-cancer rates, would not have done anything to prevent damage to crops. Plants tend to be resistant to damage from UV-B radiation, but studies show that some species can be quite sensitive.

In 1983, Alan H. Teramura of the University of Maryland reviewed the available evidence on the effects of ultraviolet radiation on crop yield and growth. He found that UV-B can interfere with photosynthesis at several critical points. Leaf area in many plants was reduced by exposure to UV-B radiation in the laboratory. Some showed leaf damage or stunted growth. A number of studies suggested that UV-B radiation might lower crop yields, but the studies were not conclusive. About 50 percent of species that have been studied are not affected by UV-B, Teramura found. About 30 percent tolerate it, and about 20 percent are sensitive even to moderate increases in UV-B.

One study found that some soybean cultivars were more sensitive to UV-B than were others. That provides an opportunity for plant breeders. The variation in sensitivity is not understood, Teramura wrote, but "it does suggest that there is a potential for genetically modifying future cultivars to minimize the deleterious effects of a global ozone depletion."[36]

As researchers continue to study the effect of UV-B on crops, the damage to the ozone layer is accelerating. In April 1991, William K. Reilly, the administrator of the EPA, warned that the depletion of the ozone layer was accelerating. New data from the National Aeronautics and Space Administration (NASA) suggested that the ozone layer was being depleted at the rate of 4 percent to 5 percent per year over the United States. "Past studies had shown about half that amount," Reilly said.

Paradoxically, while the decrease of ozone in the stratosphere is harming crops, an increase in ozone near the ground is also adversely affecting crops. Near the ground, ozone is a pollutant, not the lifesaver that it is in the upper atmosphere. EPA research suggests that ozone is the pollutant with the largest effect on crops.

The EPA has estimated that ozone pollution produced by the combustion of fossil fuels has cut U.S. corn, wheat, soybean, and peanut harvests by 5 percent.[37] The loss is put at $3 billion per year.[38]

THE LAST HARVEST

As the world's population continues to surge, the squandering and mismanagement of the world's finite supply of land, water, and energy also continue. Pesticides and fertilizers are not likely to boost yields as much in the future as they have in the past. The uncertain prospect of global warming could disrupt agriculture if the depletion of cropland and water do not disrupt it first.

The world's last harvest may be decades or even centuries

away. Or it may never occur. Agriculture is probably not going to collapse this year or next or the year after that. But with the precarious conditions associated with genetic uniformity, it could. It is a gamble as unpredictable as the wheels of a slot machine.

When it happens, breeders may not be ready. They may not have the germplasm they need. Or they will need years to get it into commercial crops because the prebreeding that would speed the process has not been done.

Glenn Fritz is one farmer in a position to do something about all of this. As much as he conforms to the traditional image of an Illinois farmer, Fritz is more politically sophisticated than he lets on. He began his political involvement on the local school board, where he served for twenty-eight years. "I wasn't satisfied with the school board. They told me to get on the school board, so I did."

Then he became involved with the Farm Bureau, a national farmers' organization. In recent years, he has become a spokesman for the group. "I've been to Washington five times in the last year," he says. Sometimes he is called at a moment's notice to help lobby Congress or the administration. He has lunched with President Bush and talked to Agriculture Secretary Mike Espy. "It's amazing how much they'll listen once you get 'em cornered," he says. Even so, he has learned that "you've got to be realistic about what you ask for." His interest in political and community affairs represents yet another break from the old ways of his father. "My dad never gave a minute to society," he says.

Once when he was in Washington he was outside a government building when he noticed indentations worn into the granite steps by the passage of countless shoes. "Imagine, granite steps worn out with shoe leather," he says reflectively, wondering how many people it must have taken to wear away the stone. "You've got to put things in perspective."

Fritz, however, is skeptical about the environmental threats to agriculture. He recognizes, for example, that pesticides have problems. He has seen one insecticide outlive its usefulness. "It got so we couldn't kill the flies," he says.

He sees hybrid crops as a necessity. "We've made the ground

grow double what it was before," he says. "When I was a lad, sixty-five bushels an acre was a terrific crop. Today, if we don't make 120 bushels, it's a disaster. We lose money at one hundred bushels an acre." In addition, he says, hybrids "stand up better in the wind. They're more drought resistant."

What if the use of hybrid crops has encouraged dangerous genetic uniformity? "It never really entered my mind. It probably wouldn't be a concern. We've always had good seed."

Global warming, again, is something he dismisses. "I think it's a lot of fuss about nothing. Statistics show you in the last few years it's gotten cooler. In 1932 and '33, we burned up. I was just a kid."

What gets Fritz agitated is something else altogether: government interference. "The government has us controlled," he says. He asks his son Gary to deal with government inspectors. "I can't tolerate 'em," Fritz says. "They don't know what they're talking about. They've read something."

That is the kind of thing that makes Fritz a little uncertain about the future. "I think America will be around for a long time. I don't know what kind of America."

If researchers can convince farmers like Fritz that environmental concerns are going to force them to change, two new technologies are on hand to help alleviate some of the problems.

One is actually an update of old technology. Alternative agriculture, or sustainable agriculture, as it is sometimes called, is a way of farming that draws inspiration from agricultural practices of the past. It seeks to reduce reliance on pesticides, for example, so it often includes crop rotations of the kind used in 1945, before the dawn of the pesticide era in American farming. Alternative agriculture is designed to be less brutal to the environment, and less destructive of water reserves and the soil.

Another technology that could help is genetic engineering. By making it possible to snip a gene from one plant and insert it into a distantly related plant or even swap genes between plants and animals, genetic engineering opens up new possibilities for plant breeders.

How these and other technologies are used will determine whether or not the world will experience Pimentel's vision of 12–15 billion people living in "absolute misery, poverty, disease, and starvation" in the year 2100.

The technologies will make new demands on the world's seed banks and its in situ preserves. Whether those demands can be met will help determine whether agriculture can delay or prevent its last harvest.

Chapter 7

Hard Science, Soft Farming: A Way Out?

In 1947, a man named J. I. Rodale bought a dilapidated sixty-eight-acre farm near Kutztown, Pennsylvania, about sixty miles north of Philadelphia, to try out some of the ideas he was espousing in his new gardening magazine. He had read a book called *An Agricultural Testament,* by an Englishman, Albert Howard, who had done research on soil. The book led Rodale to the notion that improving the soil could improve human health.

Rodale turned his farm into what he called the Soil and Health Foundation. There he tried to devise ways to grow crops without reliance on pesticides and other chemicals. He soon began urging farmers to do away with pesticides and to follow his lead.

J. I. Rodale's magazine grew into what is now the Rodale Press, the publisher of *Prevention* magazine and others, all of which have a general focus on health and fitness. His gardening magazine survived. It is now called *Organic Gardening.* He was one of the first to use the word "organic" to apply to agriculture that dispenses with chemical fertilizers and pesticides.

The Rodale farm, meanwhile, was taken over by J. I. Rodale's son, Robert, after J.I.'s death in 1971. Robert added three hundred acres to the farm and began to subject his father's ideas and experiments to more serious scientific scrutiny. The Soil and Health Foundation became the Rodale Institute, one of the leading centers for research on a new kind of agriculture that avoids heavy reliance on pesticides and fertilizers.

The new form of agriculture is usually called alternative agriculture, or sustainable agriculture. Alternative agriculture does not rely on exotic technology. It relies instead on new methods of farming—new methods of tilling the soil, new styles of crop rotation, new techniques to manage pests. The idea is, if not to eliminate pesticide use, to reduce it. At the same time, alternative agriculture aims to help cut soil erosion and conserve water.

For decades, the Rodale research has been dismissed as the work of zealots who had a kind of religious proscription against pesticides and an equally fervent appreciation of organic produce. Organic gardening was dismissed by Agriculture Department scientists as exactly that—gardening, not farming. Its methods were irrelevant to modern agriculture, they said.

Recent research suggests that the dismissal of the work of the Rodale researchers and many others who have taken up the cause was premature. Recent studies suggest that alternative agriculture's methods can reduce fertilizer and pesticide use and conserve water, soil, and energy. The research also suggests that all of this can be done at a profit.

One of the first acknowledgments of the importance of alternative agriculture by mainstream researchers came in 1989, when the National Academy of Sciences released a report that concluded that alternative agriculture could make an important contribution to American farming.

"The good news is that alternative agriculture systems and practices do work, they are environmentally beneficial, and, when effectively managed, can be highly profitable," said John Pesek, the director of the academy's study. He explained what he meant by "alternative agriculture": planting a variety of crops and rotating them; combating pests with pest-eating bugs rather than pesticides; preventing livestock disease without using antibiotics; and breeding improved crops.

The bad news in the study was that few farmers are practicing alternative agriculture. Why? "A primary reason is that numerous national policies, most notably the federal commodity programs, inhibit their development and widespread adoption," Pesek said. A more environmentally sound agricultural policy would spark growth, not hinder it.

Before the academy's study began to turn things around for alternative agriculture, advocates of organic farming were about as welcome at the Agriculture Department as boll weevils in a cotton field. The prime objective of research at the Agriculture Department had been production—getting the highest possible yields, said Richard Amerman of the Agricultural Research Service in Beltsville, Maryland. But that strict focus on the bottom line has softened.

"There is a considerable emphasis these days on what used to be called low-input sustainable agriculture, reducing the use of pesticides and fertilizers," he said in an interview in January 1994. "I think the Rodale people are fairly objective," said Amerman. One sign of the change in attitude is that the Agricultural Research Service now stations one of its own researchers at Rodale's experimental farm.

Rodale has compromised, too. It no longer takes an all-or-none approach to pesticide use. Recognizing that many farmers may be unwilling to discard pesticides overnight, the institute is working with a network of private farmers to help them meet a lesser goal, namely, reducing pesticide use.

For a brief period under the administration of President Jimmy Carter, organic-farming research gained a toehold at the USDA. In 1980, Garth Youngberg, an Agriculture Department planner, issued a report concluding that many pests could be controlled without chemicals. "It established that organic farmers exist, that organic farming can be profitable," said Rhonda Janke of Kansas State University, who was Rodale's research director until mid-1994. "There was a lot of interest in that report, a huge amount of interest," Youngberg recalled.

The Agriculture Department's response was to eliminate his job. "Just the statement that organic farming was feasible was enough to get Garth Youngberg fired," Janke said.

The Rodale Institute's research center, established in 1972 by Robert Rodale, is the oldest and largest organic farming research center in the country. The center's scientists are collaborating with thirty scientists at universities and the USDA's Agricultural Research Center, Janke said.

One of the aims of the center is to take a broad look at the agricultural ecosystem, said Michael Sands, an animal scientist and the director of the institute's international division. "It's going from the focus on the species, whether it be corn, to a focus on total productivity and on improving the total ecosystem," he said.

"We look at agriculture as being a problem in deforestation, global warming, surface and groundwater pollution, human health, and a whole variety of things. Our initial focus was on showing that sustainable agriculture was commercially feasible. Our focus in the future is really going to be on how to improve human health and the environment through agriculture."

Soil conservation is a critical component of that research, he said. "If you have a healthy, high-quality soil, you actually reduce or eliminate water pollution—both groundwater and surface pollution—because the soil absorbs it. You hold the soil in place, so you don't have sedimentation."

In the tropics, the development of healthy soils can slow deforestation, Sands said. The failure of tropical farms is one of the most important causes of deforestation. Poor farmers burn the forest, survive for a few years on the residual fertility produced by the ash of the burned trees, and then move on, destroying another patch of forest.

The institute is located in a gently hilly area that was settled by Mennonites in the 1700s. Most of the research conducted there is devoted to organic farming—producing crops without reliance on pesticides or chemical fertilizers. The researchers take a less strict approach, however, in working with farmers in a network they have established across the country.

"We don't tell farmers what to use and what not to use," said Janke. "We're saying, okay, somebody's got to pioneer the alternatives, and that's what our niche is. And we also feel that researchers should take more risks than farmers. We get paid at the end of the week whether we get a crop or not, and farmers don't." The idea is that many of the ideas developed at the institute will prove to be profitable and will be adopted by farmers. But the institute has to demonstrate that first.

One of the components of alternative agriculture, in addition to better treatment of the soil, is a kind of pest control often referred to as integrated pest management. "You monitor the pest in question—whether it's a weed, an insect, slugs," explained Janke. "You apply a very targeted dose of pesticide. Sometimes it's a biological pesticide, sometimes it's a chemical. And so you end up reducing your total pesticide load in the environment anywhere from one spray a year to more. In principle it means reducing the amount of pesticide to only what's needed."

One of the most difficult crops to grow without pesticides is apples. The Rodale Institute established an organic apple orchard in 1980. There they have tried every known strategy to combat insects without using insecticides. They have used *Bacillus thuringiensis,* a microbe that attacks insects. That is one form of biological control—using naturally occurring enemies of pests to fight the pests.

The researchers have let chickens graze under the trees so they can eat the worms that fall off them. They caught apple maggots in traps, and in another form of biological control they have disrupted the romantic courtship rituals of a moth with what are called "pheromone disruption ties."

The disruptors look like the twist ties used to seal plastic bags, but they are dipped in a moth sexual attractant, one of a class of chemicals called pheromones. Human visitors to the orchard smell only damp soil and ripe fruit. But to the moth it is awash in a dizzying sexual perfume that leaves him reeling, confused, and ultimately unfulfilled.

The organic orchard has not been perfected yet. It is still being sprayed twice a month with insecticides to control a beetle that is resistant to all of the biological-control methods, Janke said. Still, that is an enormous improvement over commercial apple breeding. "A normal apple would probably be sprayed about twenty times with insecticides and fungicides," Janke said.

The fungicides were eliminated by turning to varieties of apples that were specially bred to be resistant to fungal attacks. It is a prime example of how better utilization of crop germplasm can be an important part of efforts to develop a new, more environmentally friendly agriculture.

Rodale researchers hope to eventually dispense with the twice-a-month spraying. "We hope to get away from that by understanding the insect more," Janke said. "Where we're going with the orchard is called a habitat-management system, where you have flowering plants growing underneath the trees that provide pollen and nectar for the beneficial insects to live."

That idea of a habitat system, or a cropping system, is crucial to research on alternative agriculture. Farmers cannot simply drop pesticides, continue everything else they are doing, and expect to prosper, Janke said.

"You can't grow corn, soybeans, corn, soybeans, and withdraw pesticides," she said. That was the flaw in research done by companies that make agricultural chemicals. "The chemical companies would run experiments and withdraw all pesticides and fertilizers and get, say, ten percent of the normal yield. And they'd say, 'Well, we're all going to starve to death without pesticides and fertilizers.' In a cropping-systems context, you can get the same yield year after year after year, and nobody's going to starve to death."

One of the research institute's longest-running experimental projects illustrates what Janke means by "cropping systems." In the ten-year experiment known as the Farming Systems Trial, the researchers replaced the conventional corn-soybeans rotation with two alternative crop rotations. One includes corn, soybeans, oats, winter wheat, and red clover and is designed to be used on farms that also have animals to provide manure. The other includes corn, soybeans, oats, winter wheat, and spring barley.

In a report released in 1992, researchers found that the corn yields in the two experimental systems were equal to the corn yield in a conventional system. During the drought of 1988, the experimental systems produced more corn than the conventional rotation.[1]

The point of the study was to show that by varying the crop rotations in ways designed to minimize pest damage, it was possible to eliminate pesticides and fertilizers altogether and still get yields approximately equal to those of conventional systems, Janke said.

Legumes and manure provided the nitrogen. Red clover was planted to help with weed control. The ground was covered more of the time than it would be in a conventional system, thus reducing soil erosion. And less nitrogen escaped in the runoff.

"This is what impresses USDA," Janke said. "We went cold turkey when we withdrew the chemicals. We did have a yield decrease before the organic nitrogen kicked in, a reduction of about twenty percent. Since '84, though, all three systems have yielded the same in corn. Since the beginning of the experiment, all three systems have yielded the same in terms of soybeans. We're getting as much wheat as our county neighbors. So nobody's going to starve to death."

Does that mean, then, that farmers in Pennsylvania could eliminate their pesticides and fertilizers tomorrow and keep the same yields? "There are a couple of things that mean farmers can't go to this next year," said Sands.

They might not have the proper harvesting equipment, for example. "If you have thirty thousand dollars invested in a sprayer, you can't make the transition right away," he said. But there's also a change in approach needed, and a bit of education, Sands said.

In conventional farming, farmers spray on a schedule supplied by the county extension agent. "The extension agent says, 'For this county and this crop, spray March 12, June 15, and August 6. And here are the mixtures you should make.' And it will work," said Sands. Alternative agriculture requires that the farmer make the decisions, that he add nutrients and pest control when they are needed, not on a predetermined schedule. "The secret is management," Sands said. "Some people call it reading the ecosystem. That takes a bit of thinking. How many weeks do I have? What's the rainfall? How does my crop look? How much nitrogen is in the soil?"

Janke said that new crop varieties play an important role in the development of these alternative agricultural systems. "A lot of universities are putting all their money into biotechnology, and unfortunately they're not rehiring plant breeders," she said. "And plant breeding has to continue; otherwise, we're not going to have new varieties."

One roadblock to wider acceptance of alternative agriculture is the complicated network of government subsidies for farming. The government support system has led to chronic oversupply and forced consumers to pay higher prices for food. It has also encouraged unsustainable practices, according to Paul Faeth and John Westra of the World Resources Institute in Washington, D.C.

"The way policy is administered virtually forces farmers to deplete the natural resources on which future supplies depend," they wrote in a 1993 report entitled *Agricultural Policy and Sustainability.* The government paid farmers an estimated $17 billion in 1993, and elevated food prices cost consumers $10 billion that year. That doesn't include the billions in indirect costs associated with degradation of the environment.[2]

Not everyone is convinced that alternative agriculture can be profitable. "At present we have a lot of anecdotes, we have success stories and failure stories all over the continuum, but we don't have a defensible and comprehensive answer," said Patrick Madden, associate director of the Agriculture Department's Sustainable Agriculture Research and Education program.

Faeth argued that conventional farming looks better than it should because comparisons with alternative agriculture generally do not take into account the environmental toll of current practices. "By excluding environmental costs, you're essentially saying that they have zero value—zero value to long-term crop yield, zero value to groundwater depletion."

GENETIC ENGINEERING

Crop breeders are remarkably adept at manipulating genes. But they have not been able to do much about shortening the time it takes to produce new varieties. Only so many generations of a crop in a breeding program can be planted each year. It takes time for seeds to grow into mature plants that can be fertilized with pollen from another variety. And it takes time for the plants that have been fertilized to grow, bear flowers, and produce seeds. Greenhouses can be used to expand the growing season, but not much can be done to speed up the growth of individual plants.

The time and expense involved in breeding new crop varieties have often kept commercial plant breeders away from seed banks. Seed companies cannot spend decades developing new crop varieties. Their customers, and their investors, demand that they come out with new crop varieties more often than that. Government and university researchers use seed-bank samples to produce breeding materials that can be handed over to seed companies for use in the production of commercial crop varieties.

The development of the new science of plant genetic engineering is changing that. Genetic engineering offers the hope of revolutionizing plant breeding. In conventional breeding, crossing two species scrambles all kinds of traits, good along with bad. To move a desirable gene from one variety or breeding line to another, researchers must make crosses for many generations, eliminating undesirable traits and reinforcing desirable ones.

Genetic engineering makes it possible to move a single gene from one variety to another—or from one species to another—in a single maneuver. The gene is chemically snipped out of one plant and plugged into another. The techniques of plant genetic engineering are in their infancy. Swapping genes is simple in theory but not yet simple in practice. But the science is advancing rapidly.

Plant genetic engineering will ultimately allow researchers to perform genetic manipulations that traditional plant breeders could not have dreamed of. Before biotechnology, researchers could transfer traits only among crop relatives that could be induced to fertilize one another. A variety of tricks were devised to make so-called wide crosses between substantially different varieties. But traditional methods could never be used to transfer a disease-resistance gene from corn to wheat, for example.

With the new techniques of genetic engineering, it is becoming possible to transfer almost any gene from one organism to another. Human genes have been transferred to bacteria. The luciferase gene that makes fireflies glow has been transferred into plants, making the plants glow faintly.

The possibilities for crops are endless. Genes from corn could be transferred to rice. Even animal genes could be inserted into plants. Not all of this is possible now. Some crops are easier to manipulate than others. Researchers face numerous technical problems in learning to insert foreign genes into crops and to make the genes work the way they are supposed to. But advances are reported daily. With the advent of genetic engineering, the possibilities, in theory, are limitless.

Biotechnology's almost magical ability to shuffle genes as easily as a deck of cards has fostered the misconception that maintaining huge libraries of crop germplasm in seed banks may no longer be important. While it is true that biotechnology vastly extends the possibilities for crop breeders, it is not true that biotechnology reduces dependence on genetic resources. Quite the opposite is true: The possibilities offered by biotechnology increase the value of the genetic raw materials that genetic engineers need to work their magic.

"It is readily assumed that the industrialized world may be insulated against the loss of natural genetic diversity by technology," said one researcher. "This is not the case now, nor will it be for at least the foreseeable future. Scientists can accomplish remarkable feats in manipulating molecules and cells, but they are utterly incapable of re-creating even the simplest forms of life in test tubes. Germplasm provides our lifeline into the future."[3]

No breakthrough in fundamental research can compensate for the loss of the genetic material crop breeders depend on. The most nimble-fingered genetic engineers cannot invent a gene any more than they can invent a new kind of plant. The best they can do is take what they find in nature and improve upon it.

Environmentalists have had mixed reactions to genetic engineering. Some have opposed it outright. Others have limited their concern to the harm that might be done by releasing newly created organisms into the environment. Pimentel recalled the story of the mongoose in Puerto Rico. Imported to wipe out rats, it succeeded only in allowing one species to be replaced by another, and it now is a primary carrier of rabies.

"Genetic-engineering technology offers many opportunities for improving agriculture and making it environmentally sound," Pimentel said. The benefits include not only higher crop yields but more nutritious fruits and vegetables, reduced reliance on pesticides and fertilizers, and better soil and water conservation.

Along with the benefits, however, is "the distinct possibility of damage from this technology," Pimentel said. "Many of the risks to the environment from the release of genetically engineered organisms are similar to the problems created by the introduction of exotic species," such as the mongoose, he said. According to Pimentel, 128 crops that were deliberately introduced into the United States with some beneficial purpose in mind have become pests or weeds.

Pimentel compares the introduction of genetic engineering to the introduction of nuclear power. "At that time, the comment from nuclear engineers was, 'there is no problem; leave us alone. . . .' Failure to exercise caution could lead to serious environmental, economic and social problems as well as loss of public credibility."[4]

Pimentel also criticized the direction biotechnology is taking. "Look at the stupid things they're doing," he said. "Most of the research with plants in biotech is on herbicide resistance. Three-quarters of the research is focused on making crops resistant to herbicides. That will mean more herbicide use, more pollution." Instead, he said, genetic engineers should be focusing on producing plants with built-in resistance to insects and diseases.

One of Pimentel's current interests is perennial grains, such as might be derived from *Zea diploperennis,* Hugh Iltis's perennial wild corn relative. "Look at the energy that goes into moving the soil each year to plant these damn annual crops. If you don't move soil, you control soil erosion, you control water runoff. It has enormous implications. That's where I'd like to see our genetic engineers focusing their talents. But there's hardly anybody doing that."

Ehrlich shared some of Pimentel's assessments. The development of plant biotechnology, he said, is "potentially some of the best news for humanity." A particularly significant advance, he said, is the ability to transfer foreign genes into crop plants "to effect improvements much more quickly than is possible with traditional breeding programs."

Genetic engineering "could also play an important role in maintaining the genetic diversity of crops, since it permits the simultaneous introduction of a given useful trait into all varieties. . . . Locally adapted varieties could be genetically enhanced while remaining in production."

The Office of Technology Assessment (OTA) said that biotechnology could help to continue the twentieth century's agricultural gains. The OTA's 1991 study looked in detail at how biotechnology could affect the seed industry.[5]

American farmers spent $3.7 billion on seeds in 1988. Many of the seeds came from small firms that distribute seeds regionally. But some are larger companies that have the resources to conduct research. The companies have not spent much on research over the years. Usually the research budget has been less than 5 percent of revenue. An exception is Pioneer Hi-Bred International, which spent 7.6 percent of its budget on research in 1989.

Many of the seed companies have been acquired by larger companies in recent years. Many of the larger firms that purchased seed companies are European corporations that make pesticides and fertilizers. They include the Swiss giant Ciba-Geigy, the British chemical company ICI, and the French company Rhone-Poulenc. The U.S. chemical company Lubrizol owns eight seed companies through its subsidiary, Agrigenetics. Some food producers have also purchased seed companies, as have Calgene and Biotechnica, two genetic-engineering firms.

By 1985, four companies supplied 64 percent of the corn seed in the United States. Thirty-eight percent of it came from Pioneer.

Large seed companies have the best opportunity to exploit plant genetic engineering, the OTA concluded. Not only do they have the financial resources, they also have the germplasm and the market. Small plant biotechnology firms have often formed alliances with larger seed companies to pursue special projects. The problem is that few companies can afford the time it takes to develop commercial products from experimental, genetically engineered crops.

More basic research is needed before genetically engineered plants become practical, the OTA concluded. Nevertheless, it said, biotechnology has the potential to increase agricultural productivity by boosting yields, lowering costs, and devising new products.

THE FUTURE

The problems with the conservation of plant genetic resources are surpassed only by the failure to conserve the gene pool for livestock. For all the same reasons that plant germplasm is disappearing, animal germplasm is disappearing, too. Domesticated breeds of livestock are becoming extinct around the world. As the National Academy of Sciences pointed out in 1993, "demands from industries and consumers for greater uniformity in animal performance and animal products" has led to a decline in "the numbers of breeds, varieties, and strains of most of the agriculturally important animal species."

A recent census by a private group concerned with preserving animal germplasm examined the situation for two hundred breeds of asses, cattle, goats, horses, sheep, and pigs in North America. The group concluded that "nearly eighty breeds have very low numbers, and some are facing extinction. About half of ass, goat, sheep, and swine breeds, one-third of cattle breeds, and nearly one-quarter of horse breeds are in need of active conservation."[6] Preserving the genetic resources in those animals may prove even more difficult than preserving plant genetic resources. There is no equivalent of seed banks for livestock germplasm—another example of the failure of conservation. Aldo Leopold's injunction to "keep every cog and wheel" before tinkering is once again being ignored.

The agricultural gene pool is shrinking. No one can confidently predict that America's food supply is on the verge of collapse. What is certain, however, is that if the depletion of the crop gene pool continues, America's and the world's agriculture must ultimately suffer catastrophic losses.

It could occur this growing season or next. It might not happen until the next century. But it is almost certain to occur. Among the less fortunate people of the world, catastrophic agricultural collapses have already taken place. The consequences are staggering. Most of the estimated 200 million people who have starved to death since 1968 were children.

The demands for food in the next century could make the situation unimaginably worse, even without catastrophic agricultural collapse. The handful of researchers who understand the stakes have so far failed to persuade the U.S. government or industry to address the genetic crisis in agriculture.

At the UN Earth Summit in June 1992, President Bush stood alone among Western leaders in refusing to sign a treaty to protect the world's biological diversity. That stance was later reversed by Bill Clinton, and the Senate swiftly ratified the treaty. But so far the accord has done little to divert the giant, grinding wheels of development from the earth's remaining islands of biodiversity. If we lose too much of that biological wealth—and no one can say how much is too much—we will face the ultimate environmental crisis. We will be unable to feed ourselves.

gene pool continues, America's and the world's agriculture must ultimately suffer catastrophic losses.

It could occur this growing season or next. It might not happen until the next century. But it is almost certain to occur. Among the less fortunate people of the world, catastrophic agricultural collapses have already taken place. The consequences are staggering. Most of the estimated 200 million people who have starved to death since 1968 were children.

The demands for food in the next century could make the situation unimaginably worse, even without catastrophic agricultural collapse. The handful of researchers who understand the stakes have so far failed to persuade the U.S. government or industry to address the genetic crisis in agriculture.

At the UN Earth Summit in June 1992, President Bush stood alone among Western leaders in refusing to sign a treaty to protect the world's biological diversity. That stance was later reversed by Bill Clinton, and the Senate swiftly ratified the treaty. But so far the accord has done little to divert the giant, grinding wheels of development from the earth's remaining islands of biodiversity. If we lose too much of that biological wealth—and no one can say how much is too much—we will face the ultimate environmental crisis. We will be unable to feed ourselves.

NOTES

INTRODUCTION

1. *Wall Street Journal,* December 3, 1992, 1.
2. National Academy of Sciences, *Managing Global Genetic Resources: Agricultural Crop Issues and Policies* (Washington, D.C.: National Academy Press, 1991), 21.
3. Aldo Leopold, *A Sand County Almanac, with Essays on Conservation from Round River* (New York: Ballantine Books, 1970), 190.

CHAPTER 1

1. Noel D. Vietmeyer, "A Wild Relative May Give Corn Perennial Genes," *Smithsonian,* December 1979, 68–76.
2. Lee Mitgang, "Perils of a Seed Hunter," Associated Press, March 28, 1989; interview with Hugh Iltis, October 20, 1993.
3. Steven C. Witt, *Briefbook: Biotechnology and Genetic Diversity (* San Francisco: California Agricultural Lands Project, 1985), 44–45.
4. Hugh H. Iltis, *Contributions from the University of Wisconsin Herbarium* (Madison; University of Wisconsin Press, 1983), 3.
5. Vietmeyer, "A Wild Relative May Give Corn Perennial Genes," 68–76.

6. Norman Myers, *A Wealth of Wild Species* (Boulder, Colo.: Westview Press, 1983), 18.
7. Robert Prescott-Allen and Christine Prescott-Allen, *Genes from the Wild* (London: Earthscan Publications Ltd., 1988), 18.
8. Myers, *A Wealth of Wild Species*, 20.
9. U.S. Department of Agriculture, Office of Public Affairs, *1990 Fact Book of Agriculture,* Miscellaneous Publication No. 1063 (Washington, D.C.: U.S. Department of Agriculture, 1991), 1–3, 57.
10. Ibid., 15–17.
11. Donald L. Plucknett, "Science and Agricultural Transformation," speech at the International Food Policy Research Institute, Washington, D.C., September 9, 1993.
12. Daniel J. Kevles, *In the Name of Eugenics* (New York: Alfred A. Knopf, 1985), 41–43.
13. National Plant Genetic Resources Board, U.S. Department of Agriculture, *Plant Germplasm: Conservation and Use* (Washington, D.C.: U.S. Department of Agriculture, 1984), 13.
14. Garrison Wilkes, "Strategies for Sustaining Crop Germplasm Preservation, Enhancement, and Use" (Washington, D.C.: Consultative Group on International Agricultural Research, 1992), 19.
15. Office of Technology Assessment, U.S. Congress, *Technologies to Maintain Biological Diversity* (Washington, D.C.: U.S. Government Printing Office, 1987), 4.
16. *1990 Fact Book of Agriculture*, 131.
17. Myers, *A Wealth of Wild Species*, 30.
18. Garrison Wilkes, "N. I. Vavilov—Patriarch of Plant Genetic Resources," *Conservation Biology* 7, no. 4 (December 1993).
19. Details of Vavilov's life come from B. M. Mednikov, "The Life and Work of Nikolai Vavilov," *Impact of Science on Society* 39, no. 2 (1989): 125–32; N. I. Vavilov, *Origin and Geography of Cultivated Plants* (Cambridge: Cambridge University Press, 1992), translator's Foreword, xiv–xvi.
20. J. G. Hawkes, *Impact of Science on Society* 40, no. 2 (1990): 99.
21. J. G. Hawkes, *The Diversity of Crop Plants* (Cambridge, Mass.: Harvard University Press, 1983), 52.
22. Suzanne Possehl, "Its Budget Slashed, Russian Seed Bank Fights for Its Life," *New York Times,* March 23, 1993, C4.
23. Gabrielle Strobel, "Seeds in Need: The Vavilov Institute," *Science News,* December 18 and 25, 1993, 416–17.
24. Viktor Dragavtsev, speech at the annual meeting of the American Association for the Advancement of Science, Boston, February 13, 1993.

Notes

CHAPTER 2

1. J. S. Quick et al., *Crop Science* 31 (1991): 50–53.

2. Walter V. Reid and Kenton R. Miller, *Keeping Options Alive* (Washington, D.C.: World Resources Institute, 1989), 59.

3. Board on Agriculture, National Research Council, *Managing Global Genetic Resources* (Washington, D.C.: National Academy Press, 1993), 9.

4. National Research Council, *Managing Global Genetic Resources* (Washington, D.C.: National Academy Press, 1991).

5. J. G. Hawkes, *The Diversity of Crop Plants* (Cambridge, Mass.: Harvard University Press, 1983), 101.

6. Jack R. Kloppenburg, Jr., ed., *Seeds and Sovereignty* (Durham, N.C.: Duke University Press, 1988), 5–6.

7. Robert Prescott-Allen and Christine Prescott-Allen, *Genes from the Wild* (London: Earthscan Publications Ltd., 1988), 92–93.

8. U.S. Department of Agriculture, Agricultural Research Service, Information Staff, press release, June 6, 1991.

9. Calestous Juma, *The Gene Hunters* (Princeton, N.J.: Princeton University Press, 1989); Steven C. Witt, *Briefbook: Biotechnology and Genetic Diversity* (San Francisco: California Agricultural Lands Project, 1985).

10. Thomas Jefferson, *Public and Private Papers* (New York: Vintage Books, 1990), 377.

11. Witt, *Briefbook,* 29.

12. Dumas Malone, *Jefferson and the Rights of Man* (Boston: Little, Brown, 1951), 121–23.

13. Jack R. Kloppenburg, Jr., *First the Seed* (New York: Cambridge University Press, 1988). 53.

14. Ibid., 56.

15. Reid and Miller, *Keeping Options Alive,* 23, 25.

16. John Maxwell Hamilton, "Survival Alliances," *Gannett Center Journal* 4, no. 3 (Summer 1990): 1.

17. Donald L. Plucknett et al., *Gene Banks and the World's Food* (Princeton, N.J.: Princeton University Press, 1987), 44.

18. Gary Paul Nabhan, *Enduring Seeds* (San Francisco: North Point Press, 1989), 95–96.

19. Robert E. Perdue, Jr., and Gudrun M. Christenson, "Plant Exploration," in *Plant Breeding Reviews,* Vol. 7, ed. Jules Janick (Portland, Ore.: Timber Press, 1989), 68.

20. Plucknett et al., *Gene Banks,* 54.

21. Isabel Shipley, *Frank N. Meyer: Plant Hunter in Asia,* Cunningham, 1984, quoted in Witt, *Briefbook,* 31.

22. Plucknett et al., *Gene Banks,* 60–66.

23. Major Goodman, "What Genetic and Germplasm Stocks Are Worth Conserving," speech at the annual meeting of the American Association for the Advancement of Science, San Francisco, January 16, 1988.
24. William W. Roath, "Evaluation and Enhancement," in Janick, ed., *Plant Breeding Reviews,* Vol. 7, 183.
25. Nabhan, *Enduring Seeds,* 85.
26. Office of Technology Assessment, "Grassroots Conservation of Biological Diversity in the United States—Background Paper #1" (Washington, D.C.: U.S. Government Printing Office, 1986), 37.
27. Paul Raeburn and Lee Mitgang, "Seeds of Conflict," Associated Press, March 27–29, 1989.
28. Joel I. Cohen et al., "Ex Situ Conservation of Plant Genetic Resources: Global Development and Environmental Concerns," *Science* 253 (August 23, 1991): 866–72.
29. Steven Eberhart, personal communication, July 1994.
30. Calvin O. Qualset, interview, February 18, 1992.
31. USDA, National Plant Genetic Resources Program Report, November 1991.

CHAPTER 3

1. Warren C. Baum, *Partners Against Hunger* (Washington, D.C.: World Bank, 1986), 7–11.
2. Quoted by Jack R. Kloppenburg, Jr., and Daniel Lee Kleinman in *Seeds and Sovereignty,* ed. Jack R. Kloppenburg, Jr. (Durham, N.C.: Duke University Press, 1988).
3. Steven C. Witt, *Briefbook: Biotechnology and Genetic Diversity* (San Francisco: California Agricultural Land Project, 1985), 45–46.
4. J. G. Hawkes, "What Are Genetic Resources and Why Should They Be Conserved?," *Impact of Science on Society,* no. 158 (1990): 97–106.
5. Donald L. Plucknett et al., *Gene Banks and the World's Food* (Princeton, N.J.: Princeton University Press, 1987), 171–85.
6. Ibid., 185.
7. *New York Times,* November 11, 1988.
8. J. Trevor Williams and Quentin Jones, "Plant Germplasm Preservation: A Global Perspective," abstract of a presentation at a symposium entitled "Biological Diversity and Germplasm Preservation—Global Imperatives," held in Beltsville, Maryland, May 9–11, 1988.
9. Plucknett et al., *Gene Banks,* 155.
10. Christine Prescott-Allen and Robert Prescott-Allen, *The First Resource* (New Haven: Yale University Press, 1986).

11. Ibid., 1.
12. Ibid., 413.
13. Ibid., 233–41.
14. David Ehrenfeld, "Why Put a Value on Biodiversity?" in *Biodiversity,* ed. E. O. Wilson (Washington, D.C.: National Academy Press, 1988), 212–16.
15. Prescott-Allen and Prescott-Allen, *First Resource,* 7.
16. Ibid., 276–78, 333–34.
17. Ibid., 279–80.
18. Robert Prescott-Allen and Christine Prescott-Allen, *Genes from the Wild* (London: Earthscan Publications Ltd., 1988), 14–19.
19. Prescott-Allen and Prescott-Allen, *First Resource,* 414.
20. D. L. Doney and E. D. Whitney, "Germplasm Collection and Preservation: A Sugarbeet Experience," abstract of a presentation at a symposium entitled "Biological Diversity and Germplasm Preservation—Global Imperatives," held in Beltsville, Maryland, May 9–11,1988.
21. H. Garrison Wilkes, "Plant Genetic Resources over Ten Thousand Years: From a Handful of Seed to the Crop-Specific Mega-Gene Banks," in Kloppenburg, ed.,*Seeds and Sovereignty,*78.
22. George G. Graham et al., "Quality Protein Maize: Digestibility and Utilization by Recovering Malnourished Infants," *Pediatrics* 83, no. 3 (March 1989).
23. Norman Myers, *A Wealth of Wild Species* (Boulder, Colo.: Westview Press, 1983), 24.
24. Witt, *Briefbook,* 36.
25. Robert E. Rhoades, "The World's Food Supply at Risk," *National Geographic,* April 1991.
26. United Nations Food and Agriculture Organization, "Extinction of Plant Species Accelerating," press release, March 23, 1992.
27. Board on Agriculture, National Research Council, National Academy of Sciences, *Managing Global Genetic Resources: Agricultural Crop Issues and Policies* (Washington, D.C.: National Academy Press, 1993), 117–18, 127–28.
28. Plucknett et al, *Gene Banks,* 93.
29. Myers,*Wealth of Wild Species,* 25–26.
30. Walter V. Reid and Kenton R. Miller, *Keeping Options Alive* (Washington, D.C.: World Resources Institute, 1989), 69.
31. Prescott-Allen and Prescott-Allen, *Genes from the Wild,* 88–91.
32. Gary Paul Nabhan, *Enduring Seeds* (San Francisco: North Point Press, 1989), 109–12.
33. Ibid., 132–33.
34. Ibid., 133.
35. Ibid., 104.

36. "A Wealth of Forest Species Is Found Underfoot," *New York Times,* July 2, 1991, C1.
37. Keystone Center, "Biological Diversity on Federal Lands," report of a Keystone policy dialogue, Keystone, Colorado, April 1991, 51–52.
38. Ibid., 57–64.
39. Ibid., 64–71.
40. Ibid., 72–78.
41. Committee on the Formation of the National Biological Survey, National Research Council (Washington, D.C.: National Academy Press, 1993).
42. Ehrenfeld, "Why Put a Value on Biodiversity," 212–16.
43. Nabhan, *Enduring Seeds,* 97.
44. Ibid.
45. Edward O. Wilson, *The Diversity of Life* (Cambridge, Mass.: Harvard University Press, 1992), 332.

CHAPTER 4

1. Cecil Woodham-Smith, *The Great Hunger* (New York: Harper & Row, 1962), 31; R. Dudley Edwards and T. Desmond Williams, eds., *The Great Famine* (New York: New York University Press, 1957), 103.
2. Robert E. Rhoades, "The Incredible Potato," *National Geographic,* May 1982, 686–87.
3. Robert Prescott-Allen and Christine Prescott-Allen, *Genes from the Wild* (London: Earthscan Publications Ltd., 1988), 20.
4. Elizabeth Culotta, "How Many Genes Had to Change to Produce Corn?" *Science* 252 (June 28, 1991): 1792.
5. H. Garrison Wilkes, "Plant Genetic Resources over Ten Thousand Years: From a Handful of Seed to the Crop-Specific Mega-Gene Banks," in *Seeds and Sovereignty,* ed. Jack R. Kloppenburg, Jr. (Durham, N.C.: Duke University Press, 1988), 71.
6. Deborah Fitzgerald, *The Business of Breeding* (Ithaca, N.Y.: Cornell University Press, 1990), 57–58.
7. Ibid., 223.
8. Jack Doyle, *Altered Harvest* (New York: Penguin Books, 1985), 7.
9. Ibid., 1–2.
10. Charles S. Levings III, "The Texas Cytoplasm of Maize: Cytoplasmic Male Sterility and Disease Susceptibility," *Science* 250 (November 16, 1990): 942–47.
11. National Academy of Sciences, *Genetic Vulnerability of Major Crops* (Washington, D.C.: National Academy Press, 1972).

12. Ibid.
13. Arthur Libow, "What's Killing the Grapevines of Napa?" *New York Times* magazine, October 17, 1993, 26–28, 59–61.
14. National Academy of Sciences, *Genetic Vulnerability,* 19.
15. U.S. General Accounting Office, "The Department of Agriculture Can Minimize the Risk of Potential Crop Failures" (Washington, D.C.: U.S. General Accounting Office, 1981), 29.
16. Christine Prescott-Allen and Robert Prescott-Allen, *The First Resource* (New Haven: Yale University Press, 1986), 458.
17. Max M. Gonzalez and Paul W. Bosland, *Diversity* 7, nos. 1 and 2 (1991): 52–53.
18. Smithsonian National Museum of Natural History, symposium on "Potatoes: Feeding Tomorrow's Global Village," October 1–2, 1991.
19. Timothy Egan, *New York Times,* February 7, 1994, A10.
20. Doyle, *Altered Harvest,* 206.
21.Warren Leary, *New York Times,* October 24, 1993, 22; Robert Greene, Associated Press, May 22, 1994; Cornell University, press release, February 10, 1993.
22. Harold Faber, "A Virulent Potato Fungus Is Killing the Northeast Crop," *New York Times,* November 12, 1994, p. 26.
23. National Research Council, National Academy of Sciences, *Managing Global Genetic Resources: Agricultural Crop Issues and Policies* (Washington, D.C.: National Academy Press, 1993).
24. Garrison Wilkes, "Crop Plants: In Situ and Genetic Vulnerability, Challenges of the 1990s," presented at the annual meeting of the American Association for the Advancement of Science, Chicago, February 1992.
25. Henry L. Shands, personal communication.
26. Wilkes, "Plant Genetic Resources," 75.

CHAPTER 5

1. Niles Eldredge, *The Miner's Canary—Unraveling the Mysteries of Extinction* (New York: Prentice Hall Press, 1991).
2. The World Resources Institute, *World Resources 1990–91* (New York: Oxford University Press, 1990).
3. Jon R. Luoma, "A Wealth of Forest Species Is Found Underfoot," *New York Times,* July 2, 1991, C1, C9.
4. Agricultural Research Service, personal communication, October 25, 1991.
5. Paul Raeburn, "Unforbidden Fruit," *American Health,* January/February 1991, 50–53.

6. Hugh H. Iltis, "Serendipity in the Exploration of Biodiversity: What Good Are Weedy Tomatoes?" reprinted in *Biodiversity,* ed. E. O. Wilson (Washington, D.C.: National Academy Press, 1988), 102–3.

7. Donald L. Plucknett et al. *Gene Banks and the World's Food* (Princeton, N.J.: Princeton University Press, 1987), 94.

8. Boyce Rensberger, *Washington Post,* January 30, 1993, A18.

9. H. Garrison Wilkes, "Germplasm Preservation: Objectives and Needs," paper presented at the Beltsville Symposium XIII, "Biotic Diversity and Germplasm Preservation: Global Imperatives," May 9–11, 1988.

10. Major M. Goodman and Fernándo Castillo-Gonzalez, "Plant Genetics: Politics and Realities," in *Forum for Applied Research and Public Policy* (Knoxville, Tenn.: Oak Ridge National Laboratory, Fall 1991), 75–76.

11. Jack R. Kloppenburg, Jr., and Daniel Lee Kleinman, "Seeds of Controversy: National Property Versus Common Heritage," in *Seeds and Sovereignty,* ed. Jack R. Kloppenburg, Jr. (Durham, N.C.: Duke University Press, 1988), 188.

12. Pat Roy Mooney, "The Law of the Seed: Another Development and Plant Genetic Resources" in *Development Dialogue* (Uppsala, Sweden: Dag Hammarskjold Foundation, 1983), 4–5.

13. Goodman and Castillo-Gonzalez, *Applied Research,* 76–79.

14. John H. Barton, "Patenting Life," *Scientific American* 264, no. 3 (March 1991): 40–46.

15. Miguel Altieri, "How Common Is Our Common Future?" *Conservation Biology* 4, no. 1 (March 1990): 102–3.

CHAPTER 6

1. Garrison Wilkes, letter to *Issues in Science and Technology,* Spring 1990, 17; Robert E. Rhoades, "The World's Food Supply at Risk," *National Geographic,* April 1991.

2. Paul R. Ehrlich and Anne H. Ehrlich, *The Population Explosion* (New York: Simon & Schuster, 1990), 9, 16, 37, 65.

3. Gretchen C. Daily and Paul R. Ehrlich, "Population, Sustainability and Earth's Carrying Capacity," *Bioscience* 42, no. 10 (November 1992): 762.

4. Paul R. Ehrlich and Anne H. Ehrlich, "The Most Overpopulated Nation," *Carrying Capacity Network Clearing House Bulletin* 2, no. 8 (October 1992), 1–7.

5. Daily and Ehrlich, "Population," 763.

6. John Bongaarts, "Can the Growing Human Population Feed Itself?" *Scientific American,* March 1994, 36–42.

7. John Steinbeck, *The Grapes of Wrath* (New York: Penguin Books, 1976), 4–5.

8. Lester R. Brown, "And Today We're Going to Talk About Biodiversity . . . That's Right, Biodiversity," in *Biodiversity,* ed. E. O. Wilson (Washington, D.C.: National Academy Press, 1988), 447.

9. Lester R. Brown, "The Changing World Food Prospect: The Nineties and Beyond," *Worldwatch Paper* 85 (October 1988) (The Worldwatch Institute, Washington, D.C.).

10. Ibid.

11. David Pimentel, "Environmental and Economic Benefits of Sustainable Agriculture," in *Socio-Economic and Policy Issues for Sustainable Farming Systems,* ed. M. G. Paoletti et al. (Padua, Italy: Cooperativa Amicizia S.r.l., 1993), 7.

12. Lester R. Brown, Christopher Flavin, and Sandra Postel, *Saving the Planet* (New York: W. W. Norton and Co., 1991), 20.

13. Worldwatch Institute, *State of the World, 1990* (New York: W. W. Norton and Co., 1990), 60–61.

14. Sandra Postel, "Water for Agriculture: Facing the Limits," *Worldwatch Paper* 93 (December 1989): 5, 7.

15. Brown, "The Changing World Food Prospect."

16. Postel, "Water for Agriculture," 16–17.

17. Ibid.

18. Pimentel, "Environmental and Economic Benefits," 8.

19. Henry Kendall and David Pimentel, "Constraints on the Expansion of the Global Food Supply," Ithaca, New York,unpublished paper provided by Pimentel, 1994, 19–20.

20. The World Resources Institute, *World Resources 1992–93* (New York: Oxford University Press, 1992), 144.

21. Lester R. Brown and Worldwatch Institute, *State of the World, 1989* (New York: W. W. Norton and Co., 1989), 52–54.

22. U.S. Environmental Protection Agency, *National Water Quality Inventory* (Washington, D.C.: U.S. Environmental Protection Agency, 1990), 7.

23. Pimentel, "Environmental and Economic Benefits," 7.

24. Michael Oppenheimer and Robert H. Boyle, *Dead Heat: The Race Against the Greenhouse Effect* (New York: Basic Books, 1990), 8–17.

25. Thomas E. Graedel and Paul J. Crutzen, "The Changing Atmosphere," *Scientific American,* September 1989, 62.

26. Brown et al., *State of the World, 1989* .

27. Ehrlich and Ehrlich, *Population Explosion,* 121.

28. National Academy of Sciences, *Policy Implications of Greenhouse Warming* (Washington, D.C.: National Academy Press, 1992), 78.

29. Ibid., 659.

30. Leslie Roberts, "Academy Panel Split on Greenhouse Adaptation," *Science* 253 (September 13, 1991): 1206.
31. National Academy of Sciences, *Greenhouse Warming,* 1.
32. Ibid., 2.
33. Stephen H. Schneider, *Global Warming: Are We Entering the Greenhouse Century?* (San Francisco: Sierra Club Books, 1989): 165–75.
34. Fakhri A. Bazzaz and Eric D. Fajer, "Plant Life in a CO_2 Rich World," *Scientific American,* January 1992, 68–74.
35. Caron Blumenthal, Snow Barlow, and Colin Wrigley, in a letter to *Nature* 347 (September 20, 1990): 235.
36. Alan H. Teramura, *Physiologia Plantarum* 58 (1983): 415–27.
37. Lester R. Brown, "Feeding Six Billion," *Worldwatch,* September–October 1989, 32–40.
38. James J. MacKenzie and Mohamed T. El-Ashry, *Ill Winds: Airborne Pollution's Toll on Trees and Crops* (Washington, D.C.: World Resources Institute, 1988), 2.

CHAPTER 7

1. Steve Peters, Rhonda Janke, and Mark Bohlke, "Rodale's Farming Systems Trial, 1986–1990" (Kutztown, Penn.: Rodale Institute Research Center, 1992).
2. Paul Faeth, ed. *Agricultural Policy and Sustainability: Case Studies from India, China, the Philippines and the United States* (Washington, D.C.; World Resources Institute, 1993), 63.
3. Jack R. Kloppenburg, Jr., ed., *Seeds and Sovereignty* (Durham, N.C.: Duke University Press, 1988), 165.
4. M. G. Paoletti and D. Pimentel, "Genetic Engineering Perspectives for an Environmentally Sound, Sustainable Agriculture," March 1994, in press.
5. U.S. Congress, Office of Technology Assessment, *Biotechnology in a Global Economy* (Washington, D.C.: U.S. Government Printing Office, 1991), 99, 107–14.
6. National Research Council, National Academy of Sciences, *Managing Global Genetic Resources: Livestock* (Washington, D.C.: National Academy Press, 1993), 1, 106. Donald E. Bixby, Carolyn J. Christman, Cynthia J. Ehrman, and D. Phillip Sponenberg, *Taking Stock* (Blacksburg, Va.: McDonald and Woodward Publishing Co., 1994), 107.

Bibliography

BOOKS

Anderson, Edgar. *Plants, Man and Life.* Berkeley: University of California Press, 1969.

Baum, Warren C. *Partners Against Hunger.* Washington, D.C.: World Bank, 1986.

Brown, Lester R., Christopher Flavin, and Sandra Postel. *Saving the Planet.* New York: W. W. Norton and Co., 1991.

Carson, Rachel. *Silent Spring.* Boston: Houghton Mifflin Co., 1962.

Doyle, Jack. *Altered Harvest: Agriculture, Genetics and the Fate of the World's Food Supply.* New York: Penguin Books, 1985.

Edwards, R. Dudley, and Desmond T. Williams, eds. *The Great Famine.* New York: New York University Press, 1957.

Ehrlich, Paul R., and Anne H. Ehrlich. *The Population Explosion.* New York: Simon and Schuster, 1990.

Eldredge, Niles. *The Miner's Canary—Unraveling the Mysteries of Extinction.* New York: Prentice Hall, 1991.

Faeth, Paul, ed. *Agricultural Policy and Sustainability: Case Studies from India, China, the Philippines and the United States.* Washington, D.C.: World Resources Institute, 1993.

Fitzgerald, Deborah. *The Business of Breeding: Hybrid Corn in Illinois 1890–1940.* Ithaca, N.Y.: Cornell University Press, 1990.

Fowler, Cary, and Pat Mooney. *Shattering: Food, Politics and the Loss of Genetic Diversity.* Tucson: University of Arizona Press, 1990.

Gore, Al. *Earth in the Balance.* Boston: Houghton Mifflin Co., 1992.

Hawkes, J. G. *The Diversity of Crop Plants.* Cambridge, Mass.: Harvard University Press, 1983.

Janick, Jules, ed. *Plant Breeding Reviews,* Vol. 7: *The National Plant Germplasm System of the United States.* Portland, Ore.: Timber Press, 1989.

Jefferson, Thomas. *Public and Private Papers.* New York: Vintage Books, 1990.

Juma, Calestous. *The Gene Hunters.* Princeton, N.J.: Princeton University Press, 1989.

Kahn, E. J., Jr. *The Staffs of Life.* Boston: Little, Brown and Co., 1985.

Kennedy, Paul. *Preparing for the Twenty-first Century.* New York: Random House, 1993.

Kevles, Daniel J. *In the Name of Eugenics.* New York: Alfred A. Knopf, 1985.

Kloppenburg, Jack R., Jr., ed. *Seeds and Sovereignty: The Use and Control of Plant Genetic Resources.* Durham, N.C.: Duke University Press, 1988.

———. *First the Seed: The Political Economy of Plant Biotechnology.* New York: Cambridge University Press, 1988.

Leopold, Aldo. *A Sand County Almanac, with Essays on Conservation from Round River.* New York: Ballantine Books, 1970.

MacKenzie, James J., and Mohamed T. El-Ashry. *Ill Winds: Airborne Pollution's Toll on Trees and Crops.* Washington, D.C.: World Resources Institute, 1988.

Malone, Dumas. *Jefferson and the Rights of Man.* Boston: Little, Brown and Co., 1951.

Myers, Norman. *A Wealth of Wild Species.* Boulder, Colo.: Westview Press, 1983.

Nabhan, Gary Paul. *Enduring Seeds: Native American Agriculture and Wild Plant Conservation.* San Francisco: North Point Press, 1989.

National Academy of Sciences. *Committee on the Formation of the National Biological Survey.* Washington, D.C.: National Academy Press, 1993.

———. *Genetic Vulnerability of Major Crops.* Washington, D.C.: National Academy Press, 1972.

———. *Managing Global Genetic Resources: Agricultural Crop Issues and Policies.* Washington, D.C.: National Academy Press, 1993.

———. *Policy Implications of Greenhouse Warming: Report of the Adaptation Panel.* Washington, D.C.: National Academy Press, 1992.

———. *Policy Implications of Greenhouse Warming: Synthesis Panel.* Washington, D.C.: National Academy Press, 1992.

National Plant Genetic Resources Board. U.S. Department of Agriculture. *Plant Germplasm: Conservation and Use.* Washington, D.C.: U.S. Department of Agriculture, 1984.

National Research Council. *Alternative Agriculture.* Washington, D.C.: National Academy Press, 1989.

———. Board on Agriculture. *Managing Global Genetic Resources: Agricultural Crop Issues and Policies.* Washington, D.C.: National Academy Press, 1993.

———. *Managing Global Genetic Resources: The U.S. National Plant Germplasm System.* Washington, D.C.: National Academy Press, 1991.

Office of Technology Assessment, U.S. Congress, *Technologies to Maintain Biological Diversity.* Washington, D.C.: U.S. Government Printing Office, 1987.

———. U.S. Congress, *Biotechnology in a Global Economy.* Washington, D.C.: U.S. Government Printing Office, 1991.

Oppenheimer, Michael, and Robert H. Boyle. *Dead Heat: The Race Against the Greenhouse Effect.* New York: Basic Books, 1990.

Paoletti, M. G., et al., eds. *Socio-Economic and Policy Issues for Sustainable Farming Systems.* Padua, Italy: Cooperativa Amicizia S.r.l., 1993.

Plucknett, Donald L., et al. *Gene Banks and the World's Food.* Princeton, N.J.: Princeton University Press, 1987.

Poehlman, John Milton. *Breeding Field Crops.* 3rd ed. New York: Van Nostrand Reinhold, 1987.

Prescott-Allen, Christine, and Robert Prescott-Allen. *The First Resource.* New Haven: Yale University Press, 1986.

Prescott-Allen, Robert, and Christine Prescott-Allen. *Genes from the Wild.* London: Earthscan Publications Ltd., 1988.

Reid, Walter V., and Kenton R. Miller. *Keeping Options Alive: The Scientific Basis for Conserving Biodiversity.* Washington, D.C.: World Resources Institute, 1989.

Schneider, Stephen H. *Global Warming: Are We Entering the Greenhouse Century?* San Francisco: Sierra Club Books, 1989.

Steinbeck, John. *The Grapes of Wrath.* New York: Penguin Books, 1976.

Thoreau, Henry D. *Faith in a Seed.* Washington, D.C.: Island Press/Shearwater Books, 1993.

U.S. Department of Agriculture. Office of Public Affairs, *1990 Fact Book of Agriculture.* Miscellaneous Publication No. 1063. Washington, D.C.: U.S. Department of Agriculture, 1991.

Vavilov, N. I. *Origin and Geography of Cultivated Plants.* Cambridge, England: Cambridge University Press, 1992.

Wilson, E. O., ed. *Biodiversity*. Washington, D.C.: National Academy Press, 1988.

———. *The Diversity of Life*. Cambridge, Mass.: Harvard University Press, 1992.

Witt, Steven C. *Briefbook: Biotechnology and Genetic Diversity*. San Francisco: California Agricultural Lands Project, 1985.

Woodham-Smith, Cecil. *The Great Hunger*. New York: Harper & Row, 1962.

World Resources Institute. *World Resources 1990–91*. New York: Oxford University Press, 1990.

———. *World Resources 1992–93*. New York: Oxford University Press, 1992.

Worldwatch Institute. *State of the World, 1990*. New York: W. W. Norton and Co., 1990.

ARTICLES

Altieri, Miguel. "How Common Is Our Common Future?" *Conservation Biology* 4, no. 1 (March 1990).

Barton, John H. "Patenting Life." *Scientific American*, March 1991.

Bazzaz, Fakhri A., and Eric D. Fajer. "Plant Life in a CO_2 Rich World." *Scientific American*, January 1992.

Bongaarts, John. "Can the Growing Human Population Feed Itself?" *Scientific American*, March 1994.

Brown, Lester R. "The Changing World Food Prospect: The Nineties and Beyond." *Worldwatch Paper* 85 (October 1988).

———. "Feeding Six Billion." *World Watch*, September–October 1989.

Cohen, Joel I., et al. "Ex Situ Conservation of Plant Genetic Resources: Global Development and Environmental Concerns." *Science* 253 (August 23, 1991).

Culotta, Elizabeth. "How Many Genes Had to Change to Produce Corn?" *Science* 252 (June 28, 1991).

Daily, Gretchen C., and Paul R. Ehrlich. "Population, Sustainability and Earth's Carrying Capacity." *Bioscience*, 42, no. 10 (November 1992).

Dragavtsev, Viktor. Speech at the annual meeting of the American Association for the Advancement of Science, Boston, February 13, 1993.

Ehrlich, Paul R., and Anne Ehrlich. "The Most Overpopulated Nation." *Carrying Capacity Network Clearing House Bulletin* 2, no. 8 (October 1992).

Gonzalez, Max M., and Paul W. Bosland. "Strategies for Stemming Genetic Erosion of *Capsicum* Germplasm in the Americas." *Diversity* 7, nos. 1 and 2 (1991).

Goodman, Major M., and Fernándo Castillo-Gonzalez. "Plant Genetics: Politics and Realities." *Forum for Applied Research and Public Policy,* Oak Ridge National Laboratory, Fall 1991.

Goodman, Major. "What Genetic and Germplasm Stocks Are Worth Conserving." Speech at the annual meeting of the American Association for the Advancement of Science, San Francisco, January 16, 1988.

Graedel, Thomas E., and Paul J. Crutzen. "The Changing Atmosphere." *Scientific American,* September 1989.

Graham, George G., et al. "Quality Protein Maize: Digestibility and Utilization by Recovering Malnourished Infants." *Pediatrics* 83, no. 3 (March 1989).

Hamilton, John Maxwell. "Survival Alliances." *Gannett Center Journal* 4, no. 3 (Summer 1990).

Hawkes, J. G. "What Are Genetic Resources and Why Should They Be Conserved?" *Impact of Science on Society* 40, no. 2 (1990).

Iltis, Hugh H. *Contributions from the University of Wisconsin Herbarium,* 1, no. 1 (University of Wisconsin, Madison, 1983).

Kendall, Henry, and David Pimentel. "Constraints on the Expansion of the Global Food Supply." Ithaca, New York, 1994.

Keystone Center. "Biological Diversity on Federal Lands." Report of a Keystone policy dialogue, Keystone, Colorado, April 1991.

Levings, III, Charles S. "The Texas Cytoplasm of Maize: Cytoplasmic Male Sterility and Disease Susceptibility." *Science,* November 16, 1990.

Libow, Arthur. "What's Killing the Grapevines of Napa?" *New York Times Magazine,* October 17, 1993.

Luoma, Jon R. "A Wealth of Forest Species Is Found Underfoot." *New York Times,* July 2, 1991.

Mednikov, B. M. "The Life and Work of Nikolai Vavilov." *Impact of Science on Society* 39, no. 2 (1989): 125–32.

Mitgang, Lee. "Perils of a Seed Hunter." Associated Press, March 28, 1989.

Mooney, Pat Roy. "The Law of the Seed: Another Development and Plant Genetic Resources." *Development Dialogue* (1983): 1–2. Dag Hammarskjold Foundation, Uppsala, Sweden.

New York Times. "A Wealth of Forest Species Is Found Underfoot," July 2, 1991.

Office of Technology Assessment. "Grassroots Conservation of Biological Diversity in the United States. Background Paper #1." Washington, D.C.: U.S. Government Printing Office, 1986.

Peters, Steve, Rhonda Janke, and Mark Bohlke. "Rodale's Farming Systems Trial, 1986–1990." Rodale Institute Research Center, Kutztown, Pennsylvania, 1992.

Plucknett, Donald L. Speech at the International Food Policy Research Institute. Washington, D.C., September 9, 1993.

Postell, Suzanne. "Its Budget Slashed, Russian Seed Bank Fights for Its Life." *New York Times,* March 23, 1993.

———. "Water for Agriculture: Facing the Limits." *Worldwatch,* December 1989.

Quick, J. S., et al. "Russian Wheat Aphid Reaction and Agronomic and Quality Traits of a Resistant Wheat." *Crop Science* 31 (1991).

Raeburn, Paul. "Unforbidden Fruit." *American Health,* January/February 1991.

Raeburn, Paul, and Lee Mitgang. "Seeds of Conflict." Associated Press, March 27–29, 1989.

Rhoades, Robert E. "The Incredible Potato." *National Geographic,* May 1982.

———. "The World's Food Supply at Risk." *National Geographic,* April 1991.

Roberts, Leslie. "Academy Panel Split on Greenhouse Adaptation." *Science* 253 (September 13, 1991).

Smithsonian Institution, National Museum of Natural History. Symposium on "Potatoes: Feeding Tomorrow's Global Village," October 1–2, 1991.

Strobel, Gabrielle. "Seeds in Need: The Vavilov Institute." *Science News* 144 (December 18 and 25, 1993).

Teramura, Alan H. *Physiologia Plantarum* 5 (1983): 415–27.

U.S. Environmental Protection Agency. *National Water Quality Inventory.* Washington, D.C.: U.S. Environmental Protection Agency, 1990.

U.S. General Accounting Office. "The Department of Agriculture Can Minimize the Risk of Potential Crop Failures." Washington, D.C.: U.S. General Accounting Office, 1981.

Vietmeyer, Noel D. "A Wild Relative May Give Corn Perennial Genes." *Smithsonian,* December 1979.

Wilkes, Garrison. "N. I. Vavilov—Patriarch of Plant Genetic Resources." *Conservation Biology* 7, no. 4 (December 1993).

———. "Strategies for Sustaining Crop Germplasm Preservation, Enhancement, and Use." Consultative Group on International Agricultural Research, October 1992.

———. "Crop Plants: In Situ and Genetic Vulnerability, Challenges of the 1990s." Presented at the annual meeting of the American Association for the Advancement of Science, Chicago, February 1992.

———. "Germplasm Preservation: Objectives and Needs." Paper presented at the Beltsville Symposium XIII, Biotic Diversity and

Bibliography

Germplasm Preservation: Global Imperatives, May 9–11, 1988.

Williams, J. Trevor, and Quentin Jones. "Plant Germplasm Preservation: A Global Perspective." Abstract of a presentation at a symposium entitled "Biological Diversity and Germplasm Preservation—Global Imperatives" held in Beltsville, Maryland, May 9–11, 1988.

Acknowledgments

This book is based on hundreds of hours of interviews over the course of the past six years. I am grateful to the busy subjects of those interviews for being so generous with their time and for freely sharing the results of their research. In particular, I'd like to thank Hugh Iltis and Garrison Wilkes, whose help was invaluable.

I'd also like to thank my editors at the Associated Press, who assigned me to do the series of articles that led to this book. I would especially like to thank Lee Mitgang, a former AP colleague and a friend, who was my coauthor on the original AP articles. His enthusiasm for the subject helped keep me going.

I was fortunate to have two capable editors at Simon & Schuster. Gary Luke did a lot to shape this book, and Dominick Anfuso helped see it to completion.

My wife, Liz, and my children, Matt, Alex, and Alicia, spent many nights and weekends without me while I was traveling and writing. I'm looking forward to spending time with them on the nights and weekends to come.

I've dedicated this book to my parents, who have been an enormous inspiration to me. When I was a child, they gave me all the books I asked for; it is nice to be able to give one back.

Ridgewood, New Jersey
November 1994

Index

Index

Index

Index